Relativistic Classical Mechanics and Electrodynamics

Relativistic Classical Mechanics and Electrodynamics

Synthesis Lectures on Engineering, Science, and Technology

Relativistic Classical Mechanics and Electrodynamics
Martin Land and Lawrence P. Horwitz
2019

Relativistic Classical Mechanics and Electrodynamics

Martin Land and Lawrence P. Horwitz

ISBN: 978-3-031-00951-8   paperback
ISBN: 978-3-031-02079-7   ebook
ISBN: 978-3-031-00151-2   hardcover

7DOI 10.1007/978-3-031-02079-7

A Publication in the Springer series
SYNTHESIS LECTURES ON ENGINEERING, SCIENCE, AND TECHNOLOGY

Lecture #1
Series ISSN
ISSN pending.

Relativistic Classical Mechanics and Electrodynamics

Martin Land
Hadassah College, Jerusalem

Lawrence P. Horwitz
Tel Aviv University, Bar Ilan University, and Ariel University

SYNTHESIS LECTURES ON ENGINEERING, SCIENCE, AND TECHNOLOGY #1

ABSTRACT

This book presents classical relativistic mechanics and electrodynamics in the Feynman-Stueckelberg event-oriented framework formalized by Horwitz and Piron. The full apparatus of classical analytical mechanics is generalized to relativistic form by replacing Galilean covariance with manifest Lorentz covariance and introducing a coordinate-independent parameter τ to play the role of Newton's universal and monotonically advancing time. Fundamental physics is described by the τ-evolution of a system point through an unconstrained 8D phase space, with mass a dynamical quantity conserved under particular interactions. Classical gauge invariance leads to an electrodynamics derived from five τ-dependent potentials described by 5D pre-Maxwell field equations. Events trace out worldlines as τ advances monotonically, inducing pre-Maxwell fields by their motions, and moving under the influence of these fields. The dynamics are governed canonically by a scalar Hamiltonian that generates evolution of a 4D block universe defined at τ to an infinitesimally close 4D block universe defined at $\tau + d\tau$. This electrodynamics, and its extension to curved space and non-Abelian gauge symmetry, is well-posed and integrable, providing a clear resolution to grandfather paradoxes. Examples include classical Coulomb scattering, electrostatics, plane waves, radiation from a simple antenna, classical pair production, classical CPT, and dynamical solutions in weak field gravitation. This classical framework will be of interest to workers in quantum theory and general relativity, as well as those interested in the classical foundations of gauge theory.

KEYWORDS

spacetime, relativistic mechanics, classical electrodynamics, electrostatics, quantum field theory

Contents

Preface

This book presents classical relativistic mechanics and describes the classical electrodynamics of relativistic particles following the approach of Stueckelberg, Horwitz, and Piron (SHP). This framework, pioneered by E. C. G. Stueckelberg in 1941 and employed by Schwinger and Feynman in the development of QED, generalizes classical analytical mechanics to relativistic form by replacing Galilean covariance with Lorentz covariance, and introducing a new coordinate-independent evolution parameter τ to play the role of Newton's postulated universal and monotonically advancing time. Fundamental physics is described by the τ-evolution of a system point through an unconstrained phase space, in which each event is represented by its covariant spacetime coordinates and velocities or momenta. The full apparatus of analytical mechanics is thus made available in a manifestly covariant form, from Lagrangian and symplectic Hamiltonian methods to Noether's theorem. This approach to relativistic classical mechanics makes SHP a convenient framework for analyzing the "paradoxes" of special relativity, and in particular provides a clear resolution to the grandfather paradox.

Making the free particle Lagrangian invariant under classical gauge transformations of the first and second kind leads to an electrodynamics derived from five τ-dependent potentials, described by 5D pre-Maxwell field equations. Individual events trace out worldline trajectories as τ advances monotonically, inducing pre-Maxwell fields by their motions, and moving under the influence of these fields. The resulting theory is thus integrable and well posed, governed canonically by a scalar Hamiltonian that generates evolution of a 4D block universe defined at τ to an infinitesimally close 4D block universe defined at $\tau + d\tau$. This electrodynamics, and its extension to curved space and non-Abelian gauge symmetry, is the most general interaction possible in an unconstrained 8D phase space. We present examples that include classical Coulomb scattering, electrostatics, plane wave solutions, and radiation from a simple antenna. Standard Maxwell theory emerges from SHP as an equilibrium limit, reached by slowing the τ-evolution to zero, or equivalently, by summing the contributions over τ at each spacetime point.

A feature of SHP not present in standard Maxwell theory is that under certain conditions, particles and fields may exchange mass dynamically, under conservation of total mass, energy, and momentum. As a result, pair processes such as electron-positron creation and annihilation are permitted in classical electrodynamics, implementing Stueckelberg's original goal. Two processes that tend to restore a particle's mass to its standard value are described, one a self-interaction along the event trajectory and the other a general result in statistical mechanics. Mass restoration of this type has been found in mathematical simulations of event trajectories.

Beyond its usefulness as an approach to electrodynamics, the theory presented in this book provides the basis for a systematic, step-by-step progression from relativistic classical mechanics to relativistic quantum mechanics, many-body theory, and quantum field theory. As an example, we discuss the correspondence of the fifth classical gauge potential to the Lorentz scalar potential used in quantum mechanical two-body problems to obtain manifestly covariant solutions for the bound state, scattering experiments, and relativistic entanglement in time. Similarly, we discuss the implications of the classical relativistic mechanics for quantum field theory.

This classical framework will thus be of interest to workers in quantum theory, as well as those interested in its foundations.

Martin Land and Lawrence P. Horwitz
December 2019

Symbols

$\mu, \nu, \lambda, \rho = 0, 1, 2, 3$	4D spacetime indices
$\alpha, \beta, \gamma, \delta = 0, 1, 2, 3, 5$	5D formal indices (skipping 4)
$\eta_{\mu\nu} = \mathrm{diag}(-1, 1, 1, 1)$	4D flat Minkowski metric
$\eta_{\alpha\beta} = \mathrm{diag}(-1, 1, 1, 1, \eta_{55})$	Formal 5D flat Minkowski metric
c	Speed of light associated with $x^0 = ct$
c_5	Speed associated with $x^5 = c_5 t$
$\{F, G\} = \dfrac{\partial F}{\partial x^\mu} \dfrac{\partial G}{\partial p_\mu} - \dfrac{\partial F}{\partial p_\mu} \dfrac{\partial G}{\partial x^\mu}$	Poisson bracket
$[F, G] = FG - GF$	Commutator bracket
$\dfrac{D\dot{x}^\mu}{D\tau} = \dfrac{d\dot{x}^\mu}{d\tau} + \Gamma^\mu_{\nu\rho}\dot{x}^\nu \dot{x}^\rho$	Absolute derivative
$\nabla_\alpha X^\beta = \dfrac{\partial X^\beta}{\partial x\partial} + X^\gamma \Gamma^\beta \gamma \alpha$	Covariant derivative
$\Gamma^\mu_{\sigma\lambda} = \frac{1}{2} g^{\mu\nu} (\partial_\sigma g_{\nu\lambda} + \partial_\lambda g_{\nu\sigma} - \partial_\nu g_{\lambda\sigma})$	Christoffel symbol
$\Phi(\tau) = \delta(\tau) - (\xi\lambda)^2 \delta''(\tau)$	Interaction kernel for electromagnetic field
λ	Parameter with dimensions of time
$\xi = \frac{1}{2} \left[1 + \left(\dfrac{c_5}{c}\right)^2 \right]$	Numerical factor
$\varphi(\tau) = \lambda \Phi^{-1}(\tau)$	Inverse function for kernel

PART I

Background

CHAPTER 1

Conceptual Approaches to Spacetime

1.1 POINT MECHANICS IN 4D SPACETIME

By one measure of success, Newtonian analytical mechanics continues to outshine the modern physics that has replaced it: the impact of its underlying physical picture on conventional notions of "reality" in the wider culture. Beyond science per se, this picture was absorbed into the foundations of Enlightenment philosophy, expanding into the modern humanities and social sciences, lending it an appearance of self-evident ordinariness. Thus, in his influential textbook *Classical Mechanics*, Herbert Goldstein introduces the physical framework—space, time, simultaneity, and mass—by writing [1, p. 1] that "these concepts will not be analyzed critically here; rather, they will be assumed as undefined terms whose meanings are familiar to the reader." This familiarity is understood to flow from everyday experience with Newtonian objects $\{\mathbf{q}_n \mid n = 1, \cdots, N\}$ defined as positions in an abstract Cartesian space $\{q_n^i(t) \mid i = 1, \cdots, 3, \ n = 1, \cdots, N\}$ of infinite extent, whose configuration develops through their functional dependence on the universal time t flowing forward uniformly. Indeed, the Newtonian picture is so central to conventional understandings of the "everyday" that more than one hundred years after Einstein's *annus mirabilis*, it is the relativistic character of the Global Positioning System (GPS) found in billions of smartphones that feels distinctly unfamiliar, and "weirdness" still seems an apt term for quantum phenomena. Moreover, it is easy to forget that much of the Newtonian worldview seemed similarly "weird" to many in Newton's day, especially the uniform linearity of time, a notion seemingly at odds with certain varieties of human experience outside the laboratory, more readily described in the language of nonuniform and cyclical flows of time.

As early as 1908, Minkowski [2, p. 34] declared that: "space and time as such must fade away into shadow, and only a kind of union of the two will maintain its reality." Although initially resistant to Minkowski's tensor formulation, Einstein's 1912 exposition of special relativity [2, p. 128] elaborates the advantages of taking the event in 4D spacetime as the fundamental object. Then, in formulating general relativity, the deconstruction of the Newtonian view of space was a crucial step, as emphasized by Einstein in his 1921 lecture at Princeton [3, pp. 2–3]. Arguing that direct experience must be the basis for physical concepts, he declared that, "the earth's crust plays such a dominant role in our daily life in judging the relative position of bodies that it has led to an abstract conception of space which certainly cannot be defended." That contemporary textbooks on relativity must still repeat Einstein's identification of the spacetime event as actual experience—superseding antique notions of infinite Euclidean space he deemed illusory—indicates not only the conceptual complexity of relativity, but also a continuing cultural disparity between modern physics and other realms of human knowledge.

In 1937, Fock [4] generalized the Newtonian picture to relativistic form by writing events in 4D Minkowski spacetime as

$$\{x_n^\mu(\tau) \mid \mu = 0, \cdots, 3, \ n = 1, \cdots, N\},$$

where $x_n^0(\tau) = ct_n(\tau)$ represents the time registered for the event on the laboratory clock. These events describe a configuration that evolves as the scalar parameter τ, identified by Fock with the proper time, advances monotonically. Writing

$$\dot{x}^\mu(\tau) = \frac{dx^\mu}{d\tau}$$

he showed that by minimizing the action

$$S = \int d\tau \left(\frac{1}{2} m \dot{x}^2 + \frac{e}{c} \dot{x}_\mu A^\mu \right) \tag{1.1}$$

for a point event in an electromagnetic potential $A^\mu(x)$, one obtains the classical relativistic equations of motion. Here and in the rest of the book we take the flat metric in 4D spacetime to be

$$\eta_{\mu\nu} = \text{diag}(-1, 1, 1, 1).$$

Fock observed that the elimination of τ in favor of t in these equations is generally difficult, but is easily accomplished for the free event satisfying $\ddot{x}^\mu = 0$ as

$$\dot{x}(\tau) = \left(\dot{x}^0(\tau), \dot{\mathbf{x}}(\tau) \right) = \left(u^0, \mathbf{u} \right) \implies \frac{d\mathbf{x}}{dt} = \frac{d\mathbf{x}/d\tau}{dt/d\tau} = \frac{\mathbf{u}}{u^0/c}$$

for constant $u = \left(u^0, \mathbf{u} \right)$. Still, Fock's generalization was not yet complete.

In the Newtonian picture, a point particle whose position is described by the 3-vector trajectory $\mathbf{x}(t)$ may follow any continuous curve. In 1941, Stueckelberg [5, 6] observed that the relativistic generalization described by Fock cannot represent all possible spacetime curves because the evolution parameter is identified with the proper time of the motion. In particular, any worldline whose time evolution reverses direction must cross the spacelike region that separates future-oriented trajectories from past-oriented trajectories. Therefore, in curves of this type the sign of $\dot{x}^2(\tau)$ will change twice and the computed proper time interval

$$ds(\tau) = \frac{1}{c} \sqrt{-\eta_{\mu\nu} dx^\mu dx^\nu} = \frac{1}{c} \sqrt{-\dot{x}^2(\tau)} \, d\tau$$

fails as a parameterization. Recognizing a physical meaning in curves of this type, Stueckelberg argued for their inclusion in relativistic mechanics, requiring the introduction of an independent evolution parameter τ, analogous to the time t in the Newtonian picture, and related to the proper time s through the dynamical relation $c^2 ds^2(\tau) = -\dot{x}^2(\tau) d\tau^2$. In this, he followed Einstein's approach, by deprecating an historical abstraction he saw as an obstruction to clear physical understanding of observed phenomena.

Stueckelberg's interest in general 4D curves can be understood from Figure 1.1 on page 5 (adapted from [5]). In his model, pair annihilation is observed in curve B when the worldline reverses its time direction, because laboratory apparatus registers two events (two points on the worldline) appearing at coordinate time $t = t_1$ but none at $t = t_2$. The event first propagates forward in t (with $\dot{x}^0 > 0$) and then propagates backward in t (with $\dot{x}^0 < 0$), continuing to earlier times while advancing in space. Stueckelberg's identification of the $\dot{x}^0 < 0$ piece of the trajectory with an antiparticle observed in the laboratory will be discussed in detail in Chapter 2.

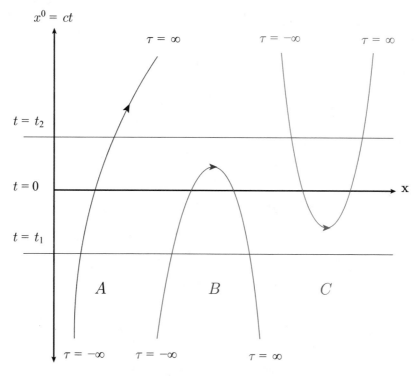

Figure 1.1: Three types of worldline identified by Stueckelberg.

In a similar way, curve C represents pair creation as two events are observed at $t = t_2$ but none at $t = t_1$. These curves may thus be seen as the smooth classical equivalent of a Feynman spacetime diagram, and the physical picture they present is known as the Feynman–Stueckelberg interpretation of antiparticles [7, 8].

Stueckelberg recognized that the standard Maxwell field $F^{\mu\nu}(x)$ alone would not permit $c^2 ds^2(\tau) = -\dot{x}^2 d\tau^2$ to change sign and proposed a modified Lorentz force

$$\frac{D\dot{x}^\mu}{D\tau} = \frac{d\,\dot{x}^\mu}{d\tau} + \Gamma^\mu_{\nu\rho}\dot{x}^\nu\dot{x}^\rho = F^{\mu\nu}(x)g_{\nu\rho}\dot{x}^\rho + G^\mu(x) \tag{1.2}$$

with local metric $g_{\mu\nu}$ and compatible connection $\Gamma^\mu_{\nu\rho}$. He also included a new vector field $G^\mu(x)$ that is required to overcome conservation of $\dot{x}^2$, as seen through

$$\frac{D}{D\tau}\dot{x}^2 = 2\dot{x}_\mu\frac{D\dot{x}^\mu}{D\tau} = 2\dot{x}_\mu G^\mu(x) \xrightarrow[G^\mu \to 0]{} 0.$$

In the absence of G^μ, spacetime curves are single-valued in x^0 and may, in principle, be reparameterized by the proper time of the motion.

As a simple example we consider a particle in flat space in a constant electric field $\mathbf{E} = E\,\hat{\mathbf{z}}$ and take $G^\mu = 0$. Writing the velocity $\dot{x} = (c\dot{t}, 0, 0, \dot{z})$, the equations of motion reduce to

$$c\ddot{t} = F^{0i}\dot{x}_i = E\dot{z} \qquad\qquad \ddot{z} = F^{30}\dot{x}_0 = cE\dot{t}$$

with solution

$$t(\tau) = \frac{1}{E}\sinh E\tau \qquad\qquad z(\tau) = \frac{c}{E}(\cosh E\tau - 1) + z(0)$$

which can be reparameterized by t as

$$z(t) = z(0) + \frac{c}{E}\left(\sqrt{1 + (Et)^2} - 1\right).$$

The velocities are

$$\dot{t}(\tau) = \cosh E\tau \qquad\qquad \dot{z}(\tau) = c\sinh E\tau$$

confirming that the mass is conserved with $\dot{x}^\mu\dot{x}_\mu = -c^2$ and so

$$c^2 = c^2\dot{t}^2 - \dot{z}^2 = c^2\left(\frac{dt}{d\tau}\right)^2\left[1 - \left(\frac{dz}{dt}\right)^2\right] \qquad\longrightarrow\qquad \dot{t} = \left[1 - \frac{v^2}{c^2}\right]^{-\frac{1}{2}},$$

where $v = dz/dt$.

Now, by contrast, we consider the particle in a constant field $G^\mu = G\,\hat{\mathbf{z}}$ and take $\mathbf{E} = 0$ so that the equations of motion are

$$c\ddot{t} = 0 \qquad\qquad \ddot{z} = G$$

with solution

$$t(\tau) = \tau \qquad\qquad z = \frac{1}{2}G\tau^2 = \frac{1}{2}Gt^2.$$

In this case the mass decreases with τ as

$$-\dot{x}^\mu\dot{x}_\mu = c^2 - G^2\tau^2$$

and the motion may become spacelike (superluminal).

1.2 THE TWO ASPECTS OF TIME

As seen in the previous section, curves B and C in Figure 1.1 cannot be parameterized by the coordinate time because they are double-valued in x^0, and cannot be parameterized by the proper time of the motion s because s^2 becomes negative in the region of the time reversal point. Realization of the classical Feynman–Stueckelberg picture thus requires the introduction of a parameter τ entirely independent of the spacetime coordinates—an irreducible chronological (historical) time, similar in its role to the external time t in nonrelativistic Newtonian mechanics. The simplicity of this picture in accounting for the observed phenomena of pair processes strongly supports the conclusion [9] that time must be understood as two distinct physical phenomena, chronology τ and coordinate x^0. A laboratory clock registers the coordinate time of an event occurrence much as a 3D array of detectors (meter sticks) registers the event's coordinate position.

The chronological time determines the order of occurrence of multiple events, with natural implications for relations of causality. Thus, when laboratory equipment reparameterizes the observed events along curve B in Figure 1.1 by x^0, two events approaching one another will be observed at $t = t_1$ and again at $t = 0$. But the underlying physics will be determined by field interactions at the locations of four distinct, ordered events, governed by a microscopic dynamics such as (1.2) and registered sequentially at $t = t_1$, $t = 0$, again at $t = 0$, and later at $t = t_1$.

In this sense, there are no closed timelike curves in this picture. The so-called grandfather paradoxes, by which one may return to an earlier time to interfere with the circumstances that brought about ones own physical presence and agency, are thus resolved. We notice that the return trip to a past coordinate time x^0 must take place while the chronological time τ continues to increase. Since the *occurrence* of event $x^\mu(\tau_1)$ at τ_1 is understood to be an irreversible process that cannot be changed by a subsequent event occurring at the same spacetime location, $x^\mu(\tau_2) = x^\mu(\tau_1)$ when $\tau_2 > \tau_1$, the return trip cannot erase the earlier trajectory. This restriction is analogous to the conceptually simpler observation in nonrelativistic physics that a process may produce new events at any given moment, but cannot delete from the historical record events that occurred at an earlier moment.

A more complex problem is the twin scenario, in which a traveler initially at rest in an inertial frame makes a trip of total distance d at speed v, so that the round trip time measured by a clock in this frame is $\Delta t = d/v$. The coordinates assigned to the traveler in the rest frame evolve as

$$
x = (ct, \mathbf{x}) = \begin{cases} (c\gamma\tau, \mathbf{x}_0 + \mathbf{u}\tau) & , \quad 0 \leq \tau \leq \Delta\tau/2 \\[2mm] (c\gamma\tau, \mathbf{x}_0 + \mathbf{u}\Delta\tau - \mathbf{u}\tau) & , \quad \Delta\tau/2 \leq \tau \leq \Delta\tau \end{cases} \quad ,
$$

where

$$
\mathbf{u} = \frac{d\mathbf{x}}{d\tau} = \frac{d\mathbf{x}}{dt}\frac{dt}{d\tau} = \gamma\mathbf{v} \qquad \Delta t = \frac{dt}{d\tau}\Delta\tau = \gamma\Delta\tau
$$

so that $d = \mathbf{u}\Delta\tau = \mathbf{v}\Delta t$. The coordinates assigned to the traveler in a co-moving frame evolve as

$$x' = (ct', \mathbf{x}') = (c\tau, \mathbf{0})$$

and so the elapsed time registered on the traveler's clock is $\Delta\tau = \Delta t/\gamma$. This result is consistent with the usual presentation of the twin scenario.

1.3   THE "PROPER TIME" FORMALISM IN QED

Although this book focuses on relativistic classical mechanics, we make a brief digression into the application of spacetime parameterization methods by Schwinger and Feynman in developing quantum electrodynamics. In his 1951 calculation of vacuum polarization in an external electromagnetic field, Schwinger [10] represented the Green's function for the Dirac field as a parametric integral and formally transformed the Dirac problem into a dynamical theory in which the integration variable acts as an independent time. Applying his method to the Klein–Gordon equation, we express the Green's function as

$$G = \frac{1}{(p - eA/c)^2 + m^2 - i\epsilon}$$

so that writing

$$G(x, x') = \langle x|G|x'\rangle = i \int_0^\infty ds e^{-i(m^2 - i\epsilon)s} \langle x|e^{-i(p - eA/c)^2 s}|x'\rangle \tag{1.3}$$

the function

$$G(x, x'; s) = \langle x(s)|x'(0)\rangle \langle x|e^{-i(p - eA/c)^2 s}|x'\rangle$$

satisfies

$$i\frac{\partial}{\partial s}\langle x(s)|x'(0)\rangle = \left(p - \frac{e}{c}A\right)^2 \langle x(s)|x'(0)\rangle \tag{1.4}$$

with the boundary condition

$$\lim_{s \to 0} \langle x(s)|x'(0)\rangle = \delta^4(x - x').$$

Schwinger regarded $x^\mu(s)$ and $\pi^\mu(s) = p^\mu(s) - \frac{e}{c}A^\mu(s)$ as operators in a Heisenberg picture that satisfy canonical relations

$$[x^\mu, \pi^\nu] = i\eta^{\mu\nu} \qquad\qquad [\pi^\mu, \pi^\nu] = \frac{ie}{c}F^{\mu\nu} \tag{1.5}$$

$$i[x^\mu, K] = -\frac{dx^\mu}{ds} \qquad\qquad i[\pi^\mu, K] = -\frac{d\pi^\mu}{ds}, \tag{1.6}$$

where $K = (p - eA/c)^2$. Using (1.5) and (1.6) we find

$$\dot{x}^{\mu}(s) = -i[x^{\mu}, K] = -i\left[x^{\mu}, \left(p - \frac{e}{c}A\right)^2\right] = 2\left(p^{\mu} - \frac{e}{c}A^{\mu}\right) \qquad (1.7)$$

and so may perform the Legendre transformation

$$\int ds\, L = \int ds\, (\dot{x}^{\mu}p_{\mu} - K) = \int ds\, \left(\frac{1}{4}\dot{x}^2 + \frac{e}{c}\dot{x}\cdot A\right)$$

whose classical limit takes the form of the Fock action (1.1). Although Schwinger found this representation useful because the scalar parameter s is necessarily independent of x^{μ} and $\dot{x}^{\mu}$, and so respects Lorentz and gauge invariance, it is known [8] as the Fock-Schwinger "proper time method."

DeWitt [11] regarded (1.4) as defining the Green's function for a Schrodinger equation

$$i\frac{\partial}{\partial s}\psi_s(x) = K\psi_s(x) = \left(p - \frac{e}{c}A\right)^2\psi_s(x) \qquad (1.8)$$

which he used for quantum mechanical calculations in curved spacetime. Similarly, Feynman [12] used (1.8) in his derivation of the path integral for the Klein–Gordon equation. He regarded the integration (1.3) of the Green's function with the weight $e^{-im^2 s}$ as the requirement that asymptotic solutions of the Schrödinger equation be stationary eigenstates of the mass operator $i\partial_{\tau}$. To pick the mass eigenvalue one extends the lower limit of integration in (1.3) from 0 to $-\infty$, and adds the requirement that $G(x, x'; s) = 0$ for $s < 0$. Feynman noted that this requirement, equivalent to imposing retarded causality in chronological time s, leads to the Feynman propagator $\Delta_F(x - x')$ whose causality properties in t are rather more complex. Related issues of causality arise in classical relativistic field theory.

1.4   THE STUECKELBERG–HORWITZ–PIRON (SHP) FRAMEWORK

In 1973, Horwitz and Piron set out to systematically construct a manifestly covariant relativistic mechanics with interactions. They observed that the principal difficulties in previous efforts arose when attempting to define observables that respect *a priori* constraints associated with the presumed dynamics. For example, although it may seem natural to choose the proper time of the motion as the worldline parameterization, Stueckelberg showed that this choice prohibits a classical account of observed pair phenomena. Worse still, in the Fock–Schwinger formalism identification of s with the proper time clashes with the formulation of quantum observables, since $\sqrt{-dx^2} = \sqrt{-\dot{x}^2}\, ds$ does not commute with x^{μ}, rendering the relations (1.5) and (1.6) difficult to interpret rigorously.

A closely related question is reparameterization invariance. Although one might regard the parameter τ as arbitrary, the Fock action (1.1) is clearly not invariant under $\tau \to \tau' = f(\tau)$

because the Lagrangian is not homogeneous of first degree in the velocities. Invariance is often restored by replacing the quadratic term in the action with a first-order form such as

$$S = \int d\tau \left(mc\sqrt{-\dot{x}^2} + \frac{e}{c}\dot{x}_\mu A^\mu \right)$$

which leads to fixed particle masses

$$p_\mu = \frac{\partial L}{\partial \dot{x}^\mu} = mc\frac{\dot{x}_\mu}{\sqrt{-\dot{x}^2}} + \frac{e}{c}A_\mu \qquad \longrightarrow \qquad \left(p - \frac{e}{c}A \right)^2 = -m^2 c^2$$

and restricts the system dynamics to the timelike region by imposing $\dot{x}^2 < 0$.

Although the Fock action permits mass exchange, the mass of individual particles is fixed for interactions governed by Stueckelberg's force law (1.2) when $G^\mu = 0$. Similarly, in the Fock–Schwinger formalism (1.7) shows that $\dot{x}^2 = 4K$ and thus corresponds to a classical constant of the motion. Thus, fixed mass is demoted from the status of *a priori* constraint to that of *a posteriori* conservation law for appropriate interactions.

Rejecting such *a priori* restrictions, Horwitz and Piron postulate that classical particles and quantum states can be described in an unconstrained 8D phase space

$$x = (ct, \mathbf{x}) \qquad\qquad p = \left(\frac{E}{c}, \mathbf{p} \right)$$

with canonical equations

$$\dot{x}^\mu = \frac{dx^\mu}{d\tau} = \frac{\partial K}{\partial p_\mu} \qquad\qquad \dot{p}_\mu = \frac{dp_\mu}{d\tau} = -\frac{\partial K}{\partial x^\mu},$$

where K is a scalar function that determines the system dynamics and its conservation laws. This framework is seen to include Newtonian mechanics by imposing the restrictions

$$t = \tau \qquad\qquad K = H(\mathbf{x}, \mathbf{p}) - E$$

which leads to

$$\frac{dx^i}{dt} = \frac{\partial H}{\partial p_i} \qquad\qquad \frac{dp^i}{dt} = -\frac{\partial H}{\partial x_i} \qquad\qquad \frac{dE}{dt} = \frac{\partial H}{\partial t},$$

where $i = 1, 2, 3$.

To describe a free relativistic particle one may write

$$K = \frac{p^2}{2M} \qquad \longrightarrow \qquad \dot{x}^\mu = \frac{p^\mu}{M} \qquad \text{and} \qquad \dot{p}^\mu = 0$$

so that $dt/d\tau = E/Mc^2$ and $d\mathbf{x}/dt = \mathbf{p}c^2/E$. In particular, for a timelike particle,

$$\dot{x}^2 = \frac{p^2}{M^2} = -\frac{m^2 c^2}{M^2} = \text{constant},$$

where the dynamical quantity $m^2(\tau)$ is conserved because $\partial K/\partial \tau = 0$. Similarly, a relativistic particle in a four-potential $A^\mu(x)$ is characterized by $K = (p - \frac{e}{c}A)^2/2M$ with results comparable to the classical limit of the Fock–Schwinger system. Moreover, Horwitz and Piron considered a two-body problem with a scalar interaction characterized by the Hamiltonian

$$K = \frac{p_1^2}{2M_1} + \frac{p_2^2}{2M_2} + V(|x_1 - x_2|),$$

where

$$V(|x_1 - x_2|) = V(\rho) = \sqrt{(\mathbf{x}_1 - \mathbf{x}_2)^2 - (t_1 - t_2)^2}$$

generalizes action at a distance to action at a spacelike interval. As in nonrelativistic mechanics, the center of mass and relative motion may be separated as

$$K = \frac{P^\mu P_\mu}{2M} + \frac{p^\mu p_\mu}{2m} + V(\rho),$$

where

$$P^\mu = p_1^\mu + p_2^\mu \qquad\qquad M = M_1 + M_2$$
$$p^\mu = \left(M_2 p_1^\mu - M_1 p_2^\mu\right)/M \qquad\qquad m = M_1 M_2/M.$$

The center of mass motion is thus free, satisfying $\dot{P}^\mu = 0$. For the relative motion, one has

$$\dot{p}^\mu = -\frac{\partial K}{\partial x_\mu} = -\frac{\partial V}{\partial x_\mu} \tag{1.9}$$

in which case we may identify $-\partial V/\partial x_\mu$ with G^μ in (1.2) so that individual particle masses are no longer necessarily fixed. In this framework, Horwitz and Arshansky found relativistic generalizations for the standard central force problems, including scattering [13, 14] and bound states [15, 16]. This formulation of the relativistic two-body problem can be extended to many bodies in the context of classical gauge theory, providing the basis for the SHP approach to classical relativistic mechanics.

1.5   BIBLIOGRAPHY

[1] Goldstein, H. 1965. *Classical Mechanics*, Addison-Wesley, Reading, MA. 3

[2] Einstein, A. 1996. *Specielle Relativitätstheorie*, George Braziller, New York, English and German on facing pages. 3

[3] Einstein, A. 1956. *The Meaning of Relativity*, Princeton University Press, Princeton, NJ. DOI: 10.4324/9780203449530. 3

[4] Fock, V. 1937. *Physikalische Zeitschrift der Sowjetunion*, 12:404–425. http://www.neo-cl assical-physics.info/uploads/3/4/3/6/34363841/fock_-_wkb_and_dirac.pdf 4

[5] Stueckelberg, E. 1941. *Helvetica Physica Acta*, 14:321–322 (in French). 4, 5

[6] Stueckelberg, E. 1941. *Helvetica Physica Acta*, 14:588–594 (in French). 4

[7] Halzen, F. and Martin, A. D. 1984. *Quarks and Leptons: An Introductory Course in Modern Particle Physics*, John Wiley & Sons, New York. DOI: 10.1119/1.14146. 5

[8] Itzykson, C. and Zuber, J. B. 1980. *Quantum Field Theory*, McGraw-Hill, New York. DOI: 10.1063/1.2916419. 5, 9

[9] Horwitz, L., Arshansky, R., and Elitzur, A. 1988. *Foundations of Physics*, 18:1159. 7

[10] Schwinger, J. 1951. *Physical Review*, 82(5):664–679. https://link.aps.org/doi/10.1103/PhysRev.82.664 8

[11] DeWitt, B. 1965. *Dynamical Theory of Groups and Fields*, Gordon and Breach, New York. DOI: 10.1119/1.1953053. 9

[12] Feynman, R. 1950. *Physical Review*, 80:440–457. 9

[13] Arshansky, R. and Horwitz, L. 1989. *Journal of Mathematical Physics*, 30:213. 11

[14] Arshansky, R. and Horwitz, L. 1988. *Physics Letter A*, 131:222–226. 11

[15] Arshansky, R. and Horwitz, L. 1989. *Journal of Mathematical Physics*, 30:66. 11

[16] Arshansky, R. and Horwitz, L. 1989. *Journal of Mathematical Physics*, 30:380. 11

PART II

Theory

CHAPTER 2

Canonical Relativistic Mechanics

2.1   LAGRANGIAN AND HAMILTONIAN MECHANICS

In many ways, the picture underlying classical relativistic mechanics is a generalization of its Newtonian predecessor, with the replacements

$$\left.\begin{array}{c} \text{3D space} \\ \text{chronological time } t \\ \text{Galilean covariance} \end{array}\right\} \longrightarrow \left\{\begin{array}{c} \text{4D spacetime} \\ \text{chronological time } \tau \\ \text{Lorentz covariance} \end{array}\right.$$

made in an analogous canonical structure. A spacetime event x^μ refers to the 4-tuple $(ct, \mathbf{x})$ of coordinate observables that can, in principle, be measured by a clock and an array of spatially arranged detectors in a laboratory.[1] Each event occurs at a chronological time τ such that for $\tau_2 > \tau_1$ the event $x^\mu(\tau_2)$ is said to occur after the event $x^\mu(\tau_1)$. Event occurrence is an irreversible process—a given event cannot be influenced by a subsequent event, although laboratory equipment may present the history of events in the order of their recorded values of $x^0 = ct$.

Following Fock and Stueckelberg, we consider a relativistic particle to be a continuous sequence of events traced out by the evolution of a function $x^\mu(\tau)$ as τ proceeds monotonically from $-\infty$ to ∞. The chronological time τ is taken to be an external universal parameter, playing a role similar to that of t in Newtonian physics.

In Stueckelberg–Horwitz–Piron (SHP) theory, event dynamics are defined on an unconstrained 8D phase space (x^μ, p^μ) by the canonical equations

$$\dot{x}^\mu = \frac{dx^\mu}{d\tau} = \frac{\partial K}{\partial p_\mu} \qquad \dot{p}^\mu = \frac{dp^\mu}{d\tau} = -\frac{\partial K}{\partial x_\mu}, \qquad (2.1)$$

where $K(x, p, \tau)$ is a Lorentz invariant Hamiltonian. This framework thus inherits the canonical structure of Newtonian analytical mechanics, with the additional complexity of Lorentz covariance. Defining Poisson brackets as

$$\{F, G\} = \frac{\partial F}{\partial x^\mu} \frac{\partial G}{\partial p_\mu} - \frac{\partial F}{\partial p_\mu} \frac{\partial G}{\partial x^\mu}$$

---

[1] Although this description oversimplifies the measurement process, it will be sufficient here.

we have for any function on phase space,

$$\frac{dF}{d\tau} = \frac{\partial F}{\partial x^\mu}\frac{dx^\mu}{d\tau} + \frac{\partial F}{\partial p_\mu}\frac{dp_\mu}{d\tau} + \frac{\partial F}{\partial \tau} = \frac{\partial F}{\partial x^\mu}\frac{\partial K}{\partial p_\mu} - \frac{\partial F}{\partial p_\mu}\frac{\partial K}{\partial x_\mu} + \frac{\partial F}{\partial \tau} = \{F, K\} + \frac{\partial F}{\partial \tau}$$

generalizing the result in nonrelativistic mechanics. Since $\{K, K\} \equiv 0$, the Hamiltonian is a constant of the motion unless K depends explicitly on τ. Because of its unconstrained canonical structure, the conditions for the Liouville–Arnold theorem apply: the 4D system is integrable—solvable by quadratures—if it possesses 8 independent conserved quantities $F_i, i = 1, \cdots, 8$ satisfying $\{K, F_i\} = 0$ and $\{F_i, F_j\} = 0$.

Performing the Legendre transformation from the Hamiltonian to the Lagrangian

$$L = \dot{x}^\mu p_\mu - K,$$

variation of the action

$$\delta S = \delta \int d\tau \, L(x, \dot{x}, \tau) = 0$$

leads to the Euler–Lagrange equations

$$\frac{d}{d\tau}\frac{\partial L}{\partial \dot{x}^\mu} - \frac{\partial L}{\partial x^\mu} = 0$$

in familiar form. Under transformations $x \to x' = f(x)$ that leave the action invariant, the Noether theorem follows in the usual manner, so that for infinitesimal variation δx we find

$$\frac{d}{d\tau}\left(\frac{\partial L}{\partial \dot{x}^\mu}\delta x^\mu\right) = 0$$

leading to the conserved quantity $(\partial L/\partial \dot{x}^\mu)\,\delta x^\mu$. In particular, since L is a scalar invariant under Lorentz transformations Λ with antisymmetric generators $M_{\mu\nu}$

$$x' = \Lambda x \qquad \longrightarrow \qquad \delta x = x' - x \simeq \delta\omega^{\mu\nu} M_{\mu\nu} x$$

the quantity

$$l_{\mu\nu} = \left(\partial L/\partial \dot{x}^\lambda\right)\left(M_{\mu\nu}\right)^{\lambda\sigma} x_\sigma = x_\mu p_\nu - x_\nu p_\mu$$

is conserved, and the Poisson bracket relations

$$\{l_{\mu\nu}, l_{\rho\sigma}\} = g_{\nu\rho}l_{\sigma\mu} + g_{\mu\rho}l_{\nu\sigma} + g_{\nu\sigma}l_{\mu\rho} + g_{\mu\sigma}l_{\rho\nu}$$

express the Lie algebra of the Lorentz group. The components of $l_{\mu\nu}$ can be split into

$$L_i = \epsilon_{ijk}x^j p^k \qquad\qquad A_i = x_0 p_i - x_i p_0 \qquad\qquad (2.2)$$

so that

$$l^2 = 2\left(L^2 - A^2\right) \qquad\qquad (2.3)$$

generalizes the conserved nonrelativistic total angular momentum in central force problems.

We write the velocity of a general event as

$$\dot{x}(\tau) = (ci, \dot{\mathbf{x}}) = (u^0(\tau), \mathbf{u}(\tau))$$

with no restrictions on its orientation—$\dot{x}$ may be timelike, lightlike, or spacelike. In the timelike case, an observer can boost to a co-moving frame in which

$$\dot{x}(\tau) = (ci, \mathbf{0})$$

and so by Lorentz invariance, $\dot{x}^2 = -(ci)^2$ in any instantaneous frame. Still, while $i = 1$ may be a dynamical result in the rest frame, it is not an *a priori* constraint.

2.2   THE FREE RELATIVISTIC PARTICLE

As in the earlier work of Fock and Schwinger, the free particle Hamiltonian is taken to be

$$K = \frac{p^2}{2M}$$

generalizing the nonrelativistic form. Applying the canonical equations (2.1), the equations of motion are

$$\dot{x}^\mu = \frac{\partial K}{\partial p_\mu} = \frac{p^\mu}{M} \qquad \dot{p}^\mu = -\frac{\partial K}{\partial x_\mu} = 0$$

with solution

$$x^\mu = x_0^\mu + u^\mu \tau = x_0^\mu + \frac{p^\mu}{M}\tau$$

as seen previously in Section 1.4. From $p^\mu = M\dot{x}^\mu$, a Legendre transformation leads to the free particle Lagrangian

$$L = \dot{x}^\mu p_\mu - K = \frac{1}{2}M\dot{x}^2$$

and so naturally,

$$L = \frac{1}{2}M\dot{x}^2 = K = \frac{p^2}{2M} = \text{constant.}$$

Given the absence of constraints, the sign of p^2 depends on its spacetime orientation.

Introducing the mass $m^2 = -p^2/c^2$ for a timelike event, we have $\dot{x}^2 = -m^2c^2/M^2$ and we generally take $m = M$ so that $i = m/M = 1$ in the rest frame. For this case,

$$-c^2 = \dot{\mathbf{x}}^2 - (\dot{x}^0)^2 = -c^2i^2\left[1 - \left(\frac{d\mathbf{x}}{dt}\right)^2\right] \qquad \longrightarrow \qquad i = \pm\frac{1}{\sqrt{1-\boldsymbol{\beta}^2}} = \pm\gamma,$$

where $\boldsymbol{\beta} = \mathbf{v}/c$, $\mathbf{v} = d\mathbf{x}/dt$, and γ is the usual relativistic dilation factor.

For a timelike free event evolving forward in coordinate time ($\dot{t} \geq 1$), we choose $\dot{t} = +\gamma$ and recover the standard representation of relativistic velocity:

$$\dot{x} = \left(u^0, \mathbf{u}\right) = \gamma \left(c, \mathbf{v}\right) = \left(\frac{E}{Mc}, \frac{\mathbf{p}}{M}\right),$$

where $E > 0$.

Choosing $\dot{t} = -\gamma$ produces a solution of particular interest to Stueckelberg, the timelike free event evolving backward in coordinate time ($\dot{t} \leq -1$),

$$= -\gamma \left(c, -\mathbf{v}\right) = \left(-\frac{|E|}{Mc}, \frac{\mathbf{p}}{M}\right)$$

describing a negative energy event tracing out a trajectory that when reordered by the laboratory clock describes an antiparticle.

The general solution $\dot{x}^\mu = p^\mu / M$ for a free particle can also accommodate tachyon ($p^2 > 0$) and lightlike ($p^2 = 0$) worldlines with no loss of generality.

2.3   THE RELATIVISTIC PARTICLE IN A SCALAR POTENTIAL

Adding a scalar potential $V(x)$ to the Hamiltonian

$$K = \frac{p^2}{2M} + V(x)$$

leads to the equations of motion

$$\dot{x}^\mu = \frac{\partial K}{\partial p_\mu} = \frac{p^\mu}{M} \qquad \dot{p}_\mu = -\frac{\partial K}{\partial x^\mu} = -\frac{\partial V}{\partial x^\mu}.$$

Equivalently, the Lagrangian formulation is

$$L = \frac{1}{2} M \dot{x}^2 - V(x)$$

$$\frac{d}{d\tau} \frac{\partial L}{\partial \dot{x}^\mu} - \frac{\partial L}{\partial x^\mu} = 0 \qquad \longrightarrow \qquad M \ddot{x}^\mu = -\frac{\partial V}{\partial x_\mu}.$$

As seen in (1.9), this problem may describe the reduced interaction of a two-body problem in relative coordinates.

As a simplified but suggestive model, we consider the scalar potential

$$V(x) = Ma \cdot x,$$

where a is a constant timelike vector. We choose a frame in which

$$a = (cg, 0, 0, 0) \qquad \longrightarrow \qquad V(x) = -Mcgx^0$$

providing an analogy in the time direction to the approximate nonrelativistic gravitational field close to earth. The equations of motion are

$$M\ddot{x}^\mu = -\frac{\partial V}{\partial x_\mu} = -Ma^\mu$$

becoming in this frame

$$M\ddot{x}^0 = -Mcg \qquad\qquad M\ddot{\mathbf{x}} = 0$$

with solution

$$t(\tau) = t_0 + \dot{t}_0\tau - \frac{1}{2}g\tau^2 \qquad\qquad \mathbf{x}(\tau) = \mathbf{x}_0 + \mathbf{u}_0\tau,$$

where g, t_0 and $\dot{t}_0$ are taken as positive constants. We recognize this parabolic trajectory as describing the pair annihilation process shown in curve B of Figure 1.1. For simplicity, we now take $t_0 = 0$ and $\mathbf{x}_0 = 0$. Thus, the event velocity is

$$\dot{t}(\tau) = \dot{t}_0 - g\tau \qquad\qquad \dot{\mathbf{x}}(\tau) = \mathbf{u}_0 = \text{constant}$$

and the trajectory reverses t-direction at $t_* = \dot{t}_0^2/2g$ when $\tau_* = \dot{t}_0/g$. From

$$p^\mu = \frac{\partial L}{\partial \dot{x}_\mu} = M\dot{x}^\mu \qquad \longrightarrow \qquad p^0 = \frac{E}{c} = Mc\dot{t} \qquad \longrightarrow \qquad E = Mc^2\dot{t}$$

we see that the event propagates forward in t with $E > 0$ for $\tau < \tau_*$ and backward in t with $E < 0$ for $\tau > \tau_*$. The $\tau > \tau_*$ portion of the trajectory corresponds to Stueckelberg's interpretation of an antiparticle. The velocity remains timelike except near τ_0 in the interval

$$c^2(\dot{t}_0 - g\tau)^2 - \mathbf{u}_0^2 < 0 \qquad \longrightarrow \qquad \tau_* - \frac{|\mathbf{u}_0|}{cg} < \tau < \tau_* + \frac{|\mathbf{u}_0|}{cg},$$

where it becomes spacelike (tachyonic).

The event trajectory recorded in the laboratory may be reordered according to t. Thus, at coordinate time $t = 0$, two events will be recorded, a positive energy event that occurred at $\tau = 0$ and a subsequent a negative energy event at $\tau = 2\tau_*$. From this perspective, the two pieces of the worldline appear as a pair of events approaching one another and mutually annihilating at $t = t_*$, with no events recorded with $t > t_*$.

In a similar way, taking $\dot{t}_0$ and g to be negative constants, this solution describes a pair creation process. Although this account of pair processes is not physically realistic, we will present a more accurate description in Section 4.6 using the full apparatus of classical SHP electrodynamics.

2.4 TWO-BODY PROBLEM WITH SCALAR POTENTIAL

As we showed in Section 1.4, the two-body problem with scalar interaction can be written as an equivalent one-body problem

$$K = \frac{p_1^2}{2M_1} + \frac{p_2^2}{2M_2} + V(x_1 - x_2) = \frac{P^\mu P_\mu}{2M} + \frac{p^\mu p_\mu}{2m} + V(x), \tag{2.4}$$

where the center of mass motion satisfies $\dot{P}^\mu = 0$. Arshansky [1] studied classical problems of this type (for the extension to quantum mechanics, see [2]), generalizing the standard nonrelativistic central force problems by taking

$$V(x) = V\left(\sqrt{x^2}\right) \qquad \longrightarrow \qquad V(x) = V\left(\sqrt{x^2 - c^2 t^2}\right)$$

for spacelike separations, $x^2 > 0$. Restriction to the spacelike region can be accomplished through a representation in hyperspherical coordinates of the type

$$x = \rho \begin{bmatrix} \sinh\beta \\ \cosh\beta\,\hat{\mathbf{r}} \end{bmatrix} \qquad\qquad \hat{\mathbf{r}} = \begin{bmatrix} \sin\theta\cos\phi \\ \sin\theta\sin\phi \\ \cos\theta \end{bmatrix} \qquad\qquad \hat{\mathbf{r}}^2 = 1.$$

But it was found that reasonable solutions lie in a subspace of the full spacelike region, found by choosing a spacelike unit vector n^μ and solving the equations of motion in the O(2,1)-invariant restricted space

$$x \in \{x \mid [x - (x \cdot n)n]^2 \ge 0\}$$

for which the component of x orthogonal to n is itself spacelike. Arshansky has described this as a classical case of spontaneous symmetry breaking leading to a lowering of the energy spectrum. Taking $n = (0, 0, 0, 1)$ this region has the representation

$$x = \rho \begin{bmatrix} \sin\theta\,\hat{q} \\ \cos\theta \end{bmatrix} \qquad\qquad \hat{q} = \begin{bmatrix} \sinh\beta \\ \cosh\beta\cos\phi \\ \cosh\beta\sin\phi \end{bmatrix} \qquad\qquad \hat{q}^2 = 1. \tag{2.5}$$

In addition to the O(3,1) invariant l^2 defined in (2.3), the O(2,1) invariant

$$N^2 = L_3^2 - A_1^2 - A_2^2$$

with components defined in (2.2) is also conserved and plays a role in characterizing the solutions. In these coordinates, the first integrals

$$K = \frac{p^2}{2M} + V(x) = \frac{1}{2}M\dot{\rho}^2 + \frac{l^2}{2M\rho^2} + V(\rho) = \kappa$$

which is cyclic in β and ϕ, and

$$l^2 = M^2 \rho^4 \dot{\theta}^2 + \frac{N^2}{\sin^2 \theta}$$

provide a separation of variables. As in nonrelativistic mechanics, but with an additional degree of freedom, solutions can be found from the four first-order equations

$$\dot{\beta} = 0$$

$$\dot{\phi} = 0$$

$$\dot{\rho} = \sqrt{\frac{2}{M}\left(\kappa - V(\rho) - \frac{l^2}{2M\rho^2}\right)} \qquad \longrightarrow \qquad \tau = \int \frac{d\rho}{\sqrt{\frac{2}{M}\left(\kappa - V(\rho) - \frac{l^2}{2M\rho^2}\right)}}$$

$$\dot{\theta} = \frac{1}{M\rho^2(\tau)}\sqrt{l^2 - \frac{N^2}{\sin^2 \theta}} \qquad \longrightarrow \qquad \int \frac{d\tau}{M\rho^2(\tau)} = \int \frac{d\theta}{\sqrt{l^2 - \frac{N^2}{\sin^2 \theta}}}$$

providing an example of Liouville integrability.

In the quantum case, Horwitz and Arshansky [2–4] solved the bound state problem, leading to a mass spectrum coinciding with the non-relativistic Schrodinger energy spectrum. For small excitations, the corresponding energy spectrum is that of the non-relativistic Schrodinger theory with relativistic corrections.

2.5   MANY-BODY PROBLEM AND STATISTICAL MECHANICS

The many body problem and classical and quantum statistical mechanics, along with applications to bound states, scattering, and relativistic statistical mechanics, are covered extensively in [5]. Here we provide a brief introduction to the subject as preparation for discussion of mass stabilization in Section 4.7.2.

The generalization of (2.4) to N-bodies is

$$K = \sum_{i=1}^{N} \frac{p_{i\mu} p_i{}^{\mu}}{2M_i} + V(x_1, x_2, \ldots, x_N)$$

for which case one may define center of mass coordinates

$$M = \sum_i M_i \qquad X^{\mu} = \frac{\sum_i M_i x_i^{\mu}}{M} \qquad P^{\mu} = \sum_i p_i^{\mu}$$

and relative coordinates

$$\hat{p}_i^{\mu} = p_i^{\mu} - (M_i/M)\, P^{\mu} \qquad \hat{x}_i^{\mu} = x_i^{\mu} - X^{\mu}$$

satisfying

$$\sum_i \hat{p}_i^\mu = 0 \qquad\qquad \sum_i M_i \hat{x}_i^\mu = 0$$

for the phase space. The Poisson brackets are

$$\{X^\mu, P^\nu\} = \eta^{\mu\nu} \qquad\qquad \{\hat{x}_i^\mu, \hat{p}_j^\nu\} = \eta^{\mu\nu}\left(\delta_{ij} - M_j/M\right)$$

and although the relative coordinates do not satisfy canonical Poisson bracket relations, these relations become canonical in the thermodynamic limit $N \to \infty$ for which $M_j/M \to 0$. The invariant Hamiltonian takes the form

$$K = \frac{P^\mu P_\mu}{2M} + \sum_i \frac{\hat{p}_i^\mu \hat{p}_{\mu i}}{2M_i} + V(x_1, x_2, \ldots, x_N)$$

so that for relative forces, $V(x_1, x_2, \ldots, x_N) = V(\hat{x}_1, \hat{x}_2, \ldots, \hat{x}_N)$ and the center of mass motion decouples from the interacting system. The equations of motion

$$\dot{X}^\mu = \frac{P^\mu}{M} \qquad\qquad \dot{P}^\mu = 0$$

$$\dot{\hat{x}}_i^\mu = \frac{P^\mu}{M} \qquad\qquad \dot{\hat{p}}_i^\mu = \frac{\partial K}{\partial \hat{x}_{\mu i}} = -\frac{\partial V}{\partial \hat{x}_{\mu i}}$$

are canonical in form.

In statistical mechanics, one regards the N events as elements in a relativistic Gibbs ensemble. As a generalization of the nonrelativistic formalism, we set a mass shell condition $K = \kappa$, however this is not a sufficient restriction because integration over the hyperbolic 4D phase space may run to infinity for finite $p^\mu p_\mu$. We must therefore also set an energy shell condition $\sum_i E_i = E$, where $E_i = p_i^0$ (we take $c = 1$ in this section). Fixing the energy shell is equivalent to choosing a Lorentz frame for the system relative to the measurement apparatus, without which we could not give meaning to the idea of temperature. The microcanonical ensemble of events at fixed energy is then defined as

$$\Gamma(\kappa, E) = \int d\Omega \, \delta(K - \kappa)\delta(\Sigma E_i - E),$$

where

$$d\Omega = \prod_i d^4 p_i \, d^4 x_i = d^{4N} p \, d^{4N} x$$

is the infinitesimal volume element in the phase space of the many-body system. The entropy and temperature are given by

$$S(\kappa, E) = \ln \Gamma(\kappa, E) \qquad\qquad T^{-1} = \frac{\partial S(\kappa, E)}{\partial E},$$

where we take the Boltzmann constant $k_B = 1$.

We may construct a canonical ensemble by extracting a small subensemble Γ_s from its environment Γ_b (the bath), and summing over all possible partitions of energy and mass parameter between the subensemble and the bath,

$$\Gamma(\kappa, E) = \int d\kappa' dE' \, \Gamma_b \left(\kappa - \kappa', E - E'\right) \Gamma_s \left(\kappa', E'\right),$$

where both mass and energy may be exchanged. Similarly, a grand canonical ensemble may be constructed by summing over all possible exchanges of event number and volume between the subensemble and the bath.

We return to the relativistic statistical mechanics in Section 4.7.2 to show that a particle represented as an ensemble of events possesses a mass that tends toward a stable equilibrium, even under perturbations.

2.6   BIBLIOGRAPHY

[1] Arshansky, R. 1986. The classical relativistic two-body problem and symptotic mass conservation. Tel Aviv University preprint TAUP 1479-86. 20

[2] Horwitz, L. P. 2015. *Relativistic Quantum Mechanics*, Springer, Dordrecht, Netherlands. DOI: 10.1007/978-94-017-7261-7. 20, 21

[3] Arshansky, R. and Horwitz, L. 1989. *Journal of Mathematical Physics*, 30:66.

[4] Arshansky, R. and Horwitz, L. 1989. *Journal of Mathematical Physics*, 30:380. 21

[5] Horwitz, L. P. and Arshansky, R. I. 2018. *Relativistic Many-Body Theory and Statistical Mechanics*, 2053–2571, Morgan & Claypool Publishers. http://dx.doi.org/10.1088/978-1-6817-4948-8 DOI: 10.1088/978-1-6817-4948-8. 21

CHAPTER 3

Classical Electrodynamics

3.1   CLASSICAL GAUGE TRANSFORMATIONS

Historically, classical electrodynamics proceeded from experiment to theory. The Maxwell equations (1860s) were initially posed as a summary of discoveries in the laboratory, including the Cavendish experiments in electrostatics (1770s), Coulomb's studies of electric and magnetic forces (1780s), and Faraday's work on time-varying fields (1830s). But the importance of Maxwell's mathematical theory was not fully recognized [1, p. xxv] until its prediction of electromagnetic waves traveling at the speed of light was verified by Hertz in 1888. It was the successful incorporation of optics into electrodynamics that provoked Einstein to study the spacetime symmetries underlying Maxwell theory in 1906 and led Fock to associate potential theory with gauge symmetry in 1929 [2]. Building on the success of such considerations, the Standard Model of fundamental interactions was developed by requiring invariance under more complex symmetry groups, as were the many candidates for a successor theory.

As discussed in Chapter 1, Stueckelberg recognized that the perception of a worldline as a sequence of events following dynamical laws could lead to pair annihilation processes in classical mechanics. Such worldlines moving in the positive or negative direction of the Einstein time t should be parameterized by an invariant τ, progressing monotonically in the positive direction. Horwitz and Piron generalized this notion to make the parameter τ universal, and in this way were able to study the relativistic classical dynamics of many body systems. In this chapter, we approach classical electrodynamics in a similar manner. Instead of restricting the formalism to the known features of Maxwell theory, we begin with the Lorentz invariant Lagrangian description of a free event and introduce the maximal U(1) gauge invariance applicable to the action, leading to a generalization of the Stueckelberg force law (1.2). We construct an action for the field strengths, again applying general principles of Lorentz and gauge invariance, and obtain τ-dependent Maxwell-like equations. The resulting framework can be understood as a microscopic theory of interacting events that reduces to Maxwell electrodynamics in a certain equilibrium limit. Thus, as we explore SHP theory, our points of comparison will be with the Maxwell theory we hope to generalize.

The action for a free event

$$S = \int d\tau \, L = \int d\tau \, \frac{1}{2} M g_{\mu\nu} \dot{x}^{\mu} \dot{x}^{\nu}$$

with $g_{\mu\nu}(x)$ a local metric, is invariant under the addition of a total τ-derivative

$$L \longrightarrow L + \frac{d}{d\tau}\Lambda(x,\tau) = L + \dot{x}^\mu \frac{\partial}{\partial x^\mu}\Lambda(x,\tau) + \frac{\partial}{\partial\tau}\Lambda(x,\tau) \qquad (3.1)$$

on condition that $\Lambda(x,\tau)$ vanishes at the endpoints of the action integral. In analogy to $x^0 = ct$, it is convenient to introduce the notation

$$x^5 = c_5\tau \qquad\qquad \dot{x}^5 = c_5 \qquad\qquad \partial_5 = \frac{1}{c_5}\frac{\partial}{\partial\tau}$$

and adopt the convention

$$\alpha,\beta,\gamma,\delta,\epsilon = 0,1,2,3,5 \qquad\qquad \lambda,\mu,\nu,\rho,\sigma = 0,1,2,3,$$

where we skip $\alpha = 4$ to avoid confusion with older notations for ct. In this notation (3.1) can be written in the compact form

$$L \longrightarrow L + \dot{x}^\alpha \partial_\alpha \Lambda(x,\tau)$$

suggesting a five dimensional symmetry acting as $x'^\alpha = L^\alpha{}_\beta x^\beta$. But we insist that in the presence of matter, x^μ and τ belong to vector and scalar representations of O(3,1). Still, free fields may enjoy a 5D symmetry, such as

$$L \in O(4,1) \text{ for metric }\quad \eta^{\alpha\beta} = \text{diag}(-1,1,1,1,+1)$$

$$L \in O(3,2) \text{ for metric }\quad \eta^{\alpha\beta} = \text{diag}(-1,1,1,1,-1)$$

which contain O(3,1) as a subgroup. Nevertheless, the higher symmetry does play a role in wave equations, much as nonrelativistic pressure waves satisfy

$$\left(\nabla^2 - \frac{1}{v^2}\frac{\partial^2}{\partial t^2}\right)p(x,t) = 0 \qquad\qquad v = \text{speed of sound}$$

suggesting a 4D symmetry not physically present in the theory of acoustics.

In light of (3.1), we introduce the five potentials $a_\alpha(x,\tau)$ into the Lagrangian and note that the action

$$S = \int d\tau\left(\frac{1}{2}M\dot{x}^\mu\dot{x}_\mu + \frac{e}{c}\dot{x}^\alpha a_\alpha\right) = \int d\tau\left[\frac{1}{2}M\dot{x}^\mu\dot{x}_\mu + \frac{e}{c}\left(\dot{x}^\mu a_\mu + c_5 a_5\right)\right]$$

is invariant under the 5D local gauge transformation

$$a_\alpha(x,\tau) \longrightarrow a'_\alpha(x,\tau) = a_\alpha(x,\tau) + \partial_\alpha\Lambda(x,\tau). \qquad (3.2)$$

As a brief quantum aside, we may write the canonical momentum

$$p_\mu = \frac{\partial L}{\partial\dot{x}^\mu} = M\dot{x}_\mu + \frac{e}{c}a_\mu \longrightarrow \dot{x}_\mu = \frac{1}{M}\left(p_\mu - \frac{e}{c}a_\mu\right)$$

to find the Hamiltonian

$$K = p_\mu \dot{x}^\mu - L = \frac{1}{2M} \left(p^\mu - \frac{e}{c} a^\mu \right) \left(p_\mu - \frac{e}{c} a_\mu \right) - \frac{ec_5}{c} a_5$$

showing that under (3.2) the Stueckelberg–Schrodinger equation

$$i\hbar \partial_\tau \psi(x, \tau) = K\psi(x, \tau) \longrightarrow \left(i\hbar \partial_\tau + \frac{ec_5}{c} a_5 \right) \psi(x, \tau) = \frac{1}{2M} \left(p - \frac{e}{c} a \right)^2 \psi(x, \tau)$$

enjoys the symmetry [3]

$$\psi(x, \tau) \rightarrow \exp \left[\frac{ie}{\hbar c} \Lambda(x, \tau) \right] \psi(x, \tau)$$

expressing the local U(1) gauge transformation in the familiar form introduced by Fock.

3.2   LORENTZ FORCE

To study the interaction of an event with the gauge potentials $a^\alpha(x, \tau)$, we write the Lagrangian as

$$L = \frac{1}{2} M g_{\mu\nu} \dot{x}^\mu \dot{x}^\nu + \frac{e}{c} g_{\mu\nu} \dot{x}^\mu a^\nu + \frac{ec_5}{c} a_5 \tag{3.3}$$

with local metric $g_{\mu\nu}(x)$. Applying the Euler–Lagrange derivative to the kinetic term we obtain

$$g^{\rho\sigma} \left(\frac{d}{d\tau} \frac{\partial}{\partial \dot{x}^\sigma} - \frac{\partial}{\partial x^\sigma} \right) \frac{1}{2} M g_{\mu\nu} \dot{x}^\mu \dot{x}^\nu = M \ddot{x}^\rho + M \left(g^{\rho\sigma} \frac{\partial g_{\sigma\mu}}{\partial x^\nu} - \frac{1}{2} g^{\rho\sigma} \frac{\partial g_{\mu\nu}}{\partial x^\sigma} \right) \dot{x}^\mu \dot{x}^\nu.$$

Using the symmetry of the first term in parentheses under $\mu \leftrightarrow \nu$

$$g^{\rho\sigma} \frac{\partial g_{\sigma\mu}}{\partial x^\nu} \dot{x}^\mu \dot{x}^\nu = \frac{1}{2} \left(g^{\rho\sigma} \frac{\partial g_{\sigma\mu}}{\partial x^\nu} + g^{\rho\sigma} \frac{\partial g_{\sigma\nu}}{\partial x^\mu} \right) \dot{x}^\mu \dot{x}^\nu$$

we find

$$g^{\rho\sigma} \left(\frac{d}{d\tau} \frac{\partial}{\partial \dot{x}^\sigma} - \frac{\partial}{\partial x^\sigma} \right) \frac{1}{2} M g_{\mu\nu} \dot{x}^\mu \dot{x}^\nu = M \ddot{x}^\rho + M \, \Gamma^\rho_{\mu\nu} \dot{x}^\mu \dot{x}^\nu = M \frac{D \dot{x}^\mu}{D\tau},$$

where

$$\Gamma^\rho_{\mu\nu} = \frac{1}{2} g^{\rho\sigma} \left(\frac{\partial g_{\sigma\mu}}{\partial x^\nu} + \frac{\partial g_{\sigma\nu}}{\partial x^\mu} - \frac{\partial g_{\mu\nu}}{\partial x^\sigma} \right)$$

is the standard Christoffel symbol and $D\dot{x}^\mu / D\tau$ is the absolute derivative of $\dot{x}^\mu$ along a geodesic. For the interaction term

$$\left(\frac{d}{d\tau} \frac{\partial}{\partial \dot{x}^\sigma} - \frac{\partial}{\partial x^\sigma} \right) (g_{\mu\nu} \dot{x}^\mu a^\nu + c_5 a_5) = \frac{d}{d\tau} \left(g_{\mu\nu} \frac{\partial \dot{x}^\mu}{\partial \dot{x}^\sigma} a^\nu \right) - \frac{\partial}{\partial x^\sigma} (g_{\mu\nu} \dot{x}^\mu a^\nu + c_5 a_5)$$

$$= \frac{da_\sigma}{d\tau} - \dot{x}^\mu \partial_\sigma a_\mu - c_5 \partial_\sigma a_5$$

$$= \dot{x}^\mu \left(\partial_\mu a_\sigma - \partial_\sigma a_\mu \right) + \partial_\tau a_\sigma - c_5 \partial_\sigma a_5$$

$$= \dot{x}^\mu \left(\partial_\mu a_\sigma - \partial_\sigma a_\mu \right) + \dot{x}^5 \left(\partial_5 a_\sigma - \partial_\sigma a_5 \right)$$

so that the Lorentz force is

$$M\ddot{x}^\rho + M \, \Gamma^\rho_{\mu\nu}\dot{x}^\mu\dot{x}^\nu = -\frac{e}{c} \, g^{\rho\sigma} \left(\dot{x}^\mu \left(\partial_\mu a_\sigma - \partial_\sigma a_\mu \right) + \dot{x}^5 \left(\partial_5 a_\sigma - \partial_\sigma a_5 \right) \right)$$

$$= \frac{e}{c} \, g^{\rho\sigma} \left(f_{\sigma\mu}\dot{x}^\mu + f_{\sigma 5}\dot{x}^5 \right), \tag{3.4}$$

where we have introduced

$$f_{\alpha\beta}\left(x,\tau\right) = \partial_\alpha a_\beta\left(x,\tau\right) - \partial_\beta a_\alpha\left(x,\tau\right) \tag{3.5}$$

as the gauge invariant field strength tensor. We note that (3.4) reduces to the Stueckelberg force (1.2) if we put

$$f_{\mu\nu}\left(x,\tau\right) \;\rightarrow\; F_{\mu\nu}\left(x\right) \qquad\qquad f_{\mu 5}\left(x,\tau\right) \;\rightarrow\; G_\mu\left(x\right)$$

and so may be said to generalize the Stueckelberg ansatz, for which it provides a foundational justification in gauge theory. In analogy to Maxwell theory, we may take $a^\mu = 0$ in (3.3) and approximate

$$-(ec_5/c)a_5(x,\tau) \simeq -(e/c)\phi(x) = V(x)$$

to identify the fifth potential with the scalar potential $V(x)$ used in Section 2.3.

We put the Lorentz force into a more compact form as

$$\frac{D\dot{x}^\rho}{D\tau} = M\ddot{x}^\rho + M \, \Gamma^\rho_{\mu\nu}\dot{x}^\mu\dot{x}^\nu = \frac{e}{c} \, g^{\rho\sigma} f_{\sigma\alpha}\dot{x}^\alpha \tag{3.6}$$

and notice that the index ρ runs to 3, while the index α runs to 5. The fifth equation is found by evaluating

$$\frac{D}{D\tau}\left(-\frac{1}{2}M\dot{x}^2\right) = -\dot{x}_\rho M \frac{D\dot{x}^\rho}{D\tau} = -\dot{x}_\rho \frac{e}{c} \, g^{\rho\sigma} \left(f_{\sigma\mu}\dot{x}^\mu + f_{\sigma 5}\dot{x}^5 \right) = \frac{c_5}{c} \, e f_{5\mu}\dot{x}^\mu, \tag{3.7}$$

where we used $f_{55} \equiv 0$. This expression shows that the $f_{5\mu}$ field, expressing the action of $a_5(x,\tau)$ and the τ-dependence of $a_\mu(x,\tau)$, permits the non-conservation of $\dot{x}^2$ and must play a role is classical pair processes. We will see in Section 3.6 that this non-conservation represents an exchange of mass between particles and fields, where total mass-energy-momentum of particles and fields is conserved.

Notice that the mass exchange is scaled by the factor c_5/c. As we shall see in Section 4.8, this factor is a continuous measure of the deviation of SHP electrodynamics from Maxwell theory, which is recovered in the limit $c_5/c \to 0$. We will generally take this factor to be small but finite.

3.3   FIELD DYNAMICS

To construct a dynamical action for the fields we first rewrite the interaction term as

$$\dot{x}^\alpha a_\alpha(x,\tau) \;\longrightarrow\; \dot{X}^\alpha a_\alpha(x,\tau) \;\longrightarrow\; \frac{1}{c}\int d^4x \; j^\alpha(x,\tau)a_\alpha(x,\tau),$$

where the event current

$$j^\alpha(x,\tau) = c\dot{X}^\alpha(\tau)\delta^4\left(x - X(\tau)\right) \tag{3.8}$$

is defined at each τ with support restricted to the spacetime location of the event at $x = X(\tau)$. The standard Maxwell current, representing the full worldline traced out by evolution of the event $X(\tau)$, is found from

$$J^\mu(x) = \int d\tau \; j^\mu(x,\tau) = c\int d\tau \; \dot{X}^\mu(\tau)\delta^4\left(x - X(\tau)\right) \tag{3.9}$$

as seen for example in [4, p. 612]. This integration is called concatenation [5] and can be understood as the sum at x of all events occurring at this spacetime location over τ.

The choice of kinetic term for a field theory is guided by three principles: it should be a Lorentz scalar, gauge invariant, and simple (bilinear in the fields with the lowest reasonable order of derivatives). From experience with the Maxwell theory, we first consider the electromagnetic action containing a term of the form $f^{\alpha\beta}(x,\tau)f_{\alpha\beta}(x,\tau)$ originally proposed by Saad et al. [3]. However, low-energy Coulomb scattering trajectories calculated in this theory [6] cannot be reconciled with Maxwell theory or experiment (we return to this point in Section 4.1). A satisfactory theory is found by generalizing the kinetic term so that the action takes the form [7]

$$S_{em} = \int d^4x\,d\tau \left\{ \frac{e}{c^2}\, j^\alpha(x,\tau)a_\alpha(x,\tau) - \int \frac{ds}{\lambda}\frac{1}{4c}\left[f^{\alpha\beta}(x,\tau)\Phi(\tau - s)f_{\alpha\beta}(x,s)\right] \right\},$$

where λ is a parameter with dimensions of time. This may be written more compactly as

$$S_{em} = \int d^4x\,d\tau \left\{ \frac{e}{c^2}\, j^\alpha(x,\tau)a_\alpha(x,\tau) - \frac{1}{4c} f_\Phi^{\alpha\beta}(x,\tau)\, f_{\alpha\beta}(x,\tau) \right\}, \tag{3.10}$$

where

$$f_\Phi^{\alpha\beta}(x,\tau) = \int \frac{ds}{\lambda}\, \Phi(\tau - s)f^{\alpha\beta}(x,s)$$

is a superposition of fields, non-local in τ. The field interaction kernel is chosen to be

$$\Phi(\tau) = \delta(\tau) - (\xi\lambda)^2\delta''(\tau) = \int \frac{d\kappa}{2\pi}\left[1 + (\xi\lambda\kappa)^2\right] e^{-i\kappa\tau}, \tag{3.11}$$

where the factor

$$\xi = \frac{1}{2}\left[1 + \left(\frac{c_5}{c}\right)^2\right] \tag{3.12}$$

insures that the low-energy Lorentz force agrees with Coulomb's law. Integrating by parts the term in (3.10) produced by the factor $\delta''(\tau - s)$ in (3.11),

$$\int d\tau ds \; f^{\alpha\beta}(x,\tau)\delta''(\tau - s) f_{\alpha\beta}(x,s) = -\int d\tau ds \left(\partial_\tau f^{\alpha\beta}(x,\tau)\right)\delta'(\tau - s) f_{\alpha\beta}(x,s)$$
$$= -\int d\tau \left(\partial_\tau f^{\alpha\beta}(x,\tau)\right)\partial_\tau f_{\alpha\beta}(x,\tau)$$

so that

$$S_{\text{em}} = \int d^4x d\tau \left\{\frac{e}{c^2} j^\alpha a_\alpha - \frac{1}{4c} f^{\alpha\beta} f_{\alpha\beta} + \frac{(\xi\lambda)^2}{4c} \left(\partial_\tau f^{\alpha\beta}\right)\left(\partial_\tau f_{\alpha\beta}\right)\right\} \qquad (3.13)$$

and the higher derivative in τ is seen to break the 5D symmetry of $f^{\alpha\beta} f_{\alpha\beta}$ to O(3,1), leaving the gauge invariance of $f^{\alpha\beta}$ unaffected. It remains necessary to give meaning to raising and lowering the 5-index through $f^5{}_\alpha = \eta^{55} f_{5\alpha}$. Expanding

$$f^{\alpha\beta} f_{\alpha\beta} = f^{\mu\nu} f_{\mu\nu} + 2\eta^{55} f_5{}^\mu f_{5\mu}$$

we see that we may interpret $\eta^{55} = \pm 1$ as the sign of the f_5^2 term in the action, sidestepping any necessary interpretation as an element in a 5D metric.

Variation of the electromagnetic action (3.10) with respect to the potentials $a^\alpha(x,\tau)$ leads to the field equations

$$\partial_\beta f_\Phi^{\alpha\beta}(x,\tau) = \frac{e}{c} j^\alpha(x,\tau) \qquad (3.14)$$

describing a non-local superposition of fields $f_\Phi^{\alpha\beta}(x,\tau)$ sourced by the local event current $j^\alpha(x,\tau)$. In order to remove $\Phi(\tau)$ from the LHS, we use the inverse function

$$\varphi(\tau) = \lambda\Phi^{-1}(\tau) = \lambda\int\frac{d\kappa}{2\pi}\frac{e^{-i\kappa\tau}}{1+(\xi\lambda\kappa)^2} = \frac{1}{2\xi}e^{-|\tau|/\xi\lambda} \qquad (3.15)$$

which satisfies

$$\int\frac{ds}{\lambda}\varphi(\tau - s)\Phi(s) = \delta(\tau) \qquad\qquad \int\frac{d\tau}{\lambda}\varphi(\tau) = 1. \qquad (3.16)$$

Integrating (3.14) with (3.15), we obtain

$$\partial_\beta f^{\alpha\beta}(x,\tau) = \frac{e}{c}\int ds\,\varphi(\tau - s)\,j^\alpha(x,s) = \frac{e}{c}j_\varphi^\alpha(x,\tau) \qquad (3.17)$$

which describes a local field sourced by a non-local superposition of event currents. While the event current (3.8) has sharp support at one spacetime point, the current

$$j_\varphi^\alpha(x,\tau) = c\int\frac{ds}{2\xi}e^{-|\tau - s|/\xi\lambda}\,\dot{X}^\alpha(s)\delta^4(x - X(s)) \qquad (3.18)$$

can be interpreted as the current induced by a smooth ensemble of events distributed in a neighborhood λ of a spacetime point. This interpretation is discussed further in Section 3.4.

Because the field strengths are derived from potentials, the Bianchi identity

$$\partial_\alpha f_{\beta\gamma} + \partial_\gamma f_{\alpha\beta} + \partial_\beta f_{\gamma\alpha} = 0 \tag{3.19}$$

holds. We see that (3.17) and (3.19) are formally similar to Maxwell's equations in 5D, and are known as pre-Maxwell equations.

Expanding the field equations in 4D tensor, vector and scalar components, they take the form

$$\partial_\nu f^{\mu\nu} - \frac{1}{c_5} \frac{\partial}{\partial \tau} f^{5\mu} = \frac{e}{c} j^\mu_\varphi \qquad\qquad \partial_\mu f^{5\mu} = \frac{e}{c} j^5_\varphi$$

$$\partial_\mu f_{\nu\sigma} + \partial_\nu f_{\sigma\mu} + \partial_\sigma f_{\mu\nu} = 0 \qquad\qquad \partial_\nu f_{5\mu} - \partial_\mu f_{5\nu} + \frac{1}{c_5} \frac{\partial}{\partial \tau} f_{\mu\nu} = 0 \tag{3.20}$$

which when compared with the 3-vector form of Maxwell's equations

$$\nabla \times \mathbf{B} - \frac{1}{c} \frac{\partial}{\partial t} \mathbf{E} = \frac{e}{c} \mathbf{J} \qquad\qquad \nabla \cdot \mathbf{E} = \frac{e}{c} J^0$$

$$\nabla \cdot \mathbf{B} = 0 \qquad\qquad \nabla \times \mathbf{E} + \frac{1}{c} \frac{\partial}{\partial t} \mathbf{B} = 0$$

suggest that $f^{5\mu}$ plays the role of the electric field, whose divergence provides the Gauss law, and $f^{\mu\nu}$ plays the role of the magnetic field. It follows from (3.17) that

$$\partial_\alpha j^\alpha = \partial_\mu j^\mu + \frac{1}{c_5} \frac{\partial}{\partial \tau} j^\alpha = 0 \tag{3.21}$$

so that $j^5(x, \tau) = c_5 \rho(x, \tau)$ plays the role of an event density, and

$$\frac{d}{d\tau} \int d^4x \, \rho(x, \tau) = -\int d^4x \, \partial_\mu j^\mu(x, \tau) = 0$$

shows the conservation of total event number over spacetime, in the absence of injection/removal of events at the boundary by an external process.

3.4   ENSEMBLE OF EVENT CURRENTS

The function $\varphi(\tau)$ smooths the current defined sharply at the event, over a range determined by λ. For λ very large, $\varphi(\tau) \simeq 1$ for all τ, producing a current ensemble associated with a large section of the worldline, approximating the standard Maxwell current. For $\lambda \to 0$, we approach the limit $\varphi(\tau)/\lambda \to \delta(\tau)$ which restricts the source current to the instantaneous current produced by a single event.

Rewriting the current (3.18) as

$$j_\varphi^\alpha (x, \tau) = \int ds \varphi (\tau - s) \, j^\alpha (x, s) = \frac{1}{2\xi} \int ds \, e^{-|s|/\xi\lambda} \, j^\alpha (x, \tau - s)$$

we recognize $j_\varphi^\alpha (x, \tau)$ as a weighted superposition of currents. Each of these currents originates at an event $X^\mu(\tau - s)$ along the worldline, occurring before or after the event $X^\mu(\tau)$, depending on the displacement s. The superposition may thus be seen [8] as the current produced by an ensemble of events in the neighborhood of $X^\mu(\tau)$, a probabilistic view encouraged by the functional form of the weight $\varphi(s)$. Consider a Poisson distribution describing the occurrence of independent random events produced at a constant average rate of $1/\lambda\xi$ events per second. The average time between events is $\lambda\xi$ and the probability at τ that the next event will occur following a time interval $s > 0$ is just $\varphi(s)/\lambda = e^{-s/\xi\lambda}/\xi\lambda$, which may be extended to positive and negative values of the displacement. The current $j_\varphi^\alpha (x, \tau)$ is constructed by assembling a set of event currents $j^\alpha (x, \tau - s)$ along the worldline, each weighted by $\varphi(s)$, the probability that the event occurrence is delayed from τ by an interval of at least $|s|$. We will see that the causality relations embedded in the pre-Maxwell equations select the one event from this ensemble for which an interaction occurs at lightlike separation, preserving relativistic causality.

We may also regard $j_\varphi^\alpha (x, \tau)$ as a random variable describing the probability of finding a current density at x at a given τ. The correlation function for the event density is

$$\langle \rho (\tau) \rho (s) \rangle = \frac{1}{N} \int d^4x \, \rho (x, \tau) \rho (x, s) ,$$

where N is a normalization. In the case of an event $X^\mu(\tau) = u^\mu \tau$ with constant velocity u^μ, the unsmoothed event current (3.8) leads to

$$\langle \rho (\tau) \rho (s) \rangle = \frac{c^2}{N} \int d^4x \, \delta^4 (x - u\tau) \delta^4 (x - us) = \frac{c^2 \delta^3 (0)}{|u^0| N} \delta(\tau - s)$$

showing that the currents at differing times $\tau \neq s$ are uncorrelated. For the ensemble current defined in (3.18) the correlation becomes

$$\langle \rho_\varphi (\tau) \rho_\varphi (s) \rangle = \frac{c^2}{N} \int d\tau' ds' d^4x \, \varphi(\tau - \tau')\varphi(s - s')\delta^4 (x - u\tau') \, \delta^4 (x - us')$$

$$= \frac{c^2 \delta^3 (0)}{|u^0| N} \int d\tau' \, \varphi(\tau - \tau')\varphi(\tau' - s)$$

$$= \frac{c^2 \delta^3 (0)}{4\xi^2 |u^0| N} \int d\tau' \, e^{-|\tau - \tau'|/\xi\lambda - |\tau' - s|/\xi\lambda} .$$

Taking $\tau > s$ and evaluating the integral over three intervals punctuated by s, τ', and τ leads to

$$\langle \rho_\varphi (\tau) \rho_\varphi (s) \rangle = \frac{\lambda c^2 \delta^3 (0)}{4\xi |u^0| N} \left(1 + \frac{\tau - s}{\xi\lambda}\right) e^{-(\tau - s)/\xi\lambda}$$

with a time-dependence characteristic of an Ornstein–Uhlenbeck process with correlation length λ. This correlation suggests that the current ensemble may be seen as the set of instantaneous currents induced by an event undergoing a Brownian motion that produces random displacement in τ under viscous drag along the worldline.

3.5   THE 5D WAVE EQUATION AND ITS GREEN'S FUNCTIONS

Using (3.5) to expand (3.17) leads to the wave equation

$$- \partial_\beta f^{\alpha\beta} = -\partial_\beta \left(\partial^\alpha a^\beta - \partial^\beta a^\alpha \right) = \partial_\beta \partial^\beta a^\alpha = \left(\partial_\mu \partial^\mu + \frac{\eta_{55}}{c_5^2} \partial_\tau^2 \right) a^\alpha = -\frac{e}{c} j_\varphi^\alpha, \qquad (3.22)$$

where we work in the 5D Lorenz gauge $\partial_\beta a^\beta = 0$. As discussed above, this form partially preserves 5D symmetries broken by the O(3,1) symmetry of the event dynamics. A Green's function solution to

$$\left(\partial_\mu \partial^\mu + \frac{\eta_{55}}{c_5^2} \partial_\tau^2 \right) G(x, \tau) = -\delta^4(x) \delta(\tau)$$

can be used to obtain potentials in the form

$$a^\alpha(x, \tau) = -\frac{e}{c} \int d^4 x' d\tau' \, G\left(x - x', \tau - \tau' \right) j_\varphi^\alpha \left(x', \tau' \right) . \qquad (3.23)$$

The Green's function can be expressed as the Fourier transform

$$G(x, \tau) = \frac{1}{(2\pi)^5} \int_C d^5 k \, \frac{e^{ik_\alpha x^\alpha}}{k^\alpha k_\alpha} = \frac{1}{(2\pi)^5} \int_C d^4 k \, d\kappa \, e^{i(k \cdot x + c_5 \eta_{55} \kappa \tau)} \frac{1}{k^2 + \eta_{55} \kappa^2}$$

over an appropriate contour C. To break the 5D symmetry present in the wave equation, we leave the κ integration for last, writing

$$G(x, \tau) = \frac{1}{2\pi} \int d\kappa \, e^{ic_5 \eta_{55} \kappa \tau} \Delta\left(x, \eta_{55} \kappa^2 \right),$$

where $\Delta(x, m^2)$ is Schwinger's principal part Green's function [9] associated with the Klein–Gordon equation for a particle of mass m. Carefully repeating the steps of Schwinger's derivation, while allowing η_{55} to be positive or negative, we are led to

$$G(x, \tau) = -\frac{1}{(2\pi)^2} \int d\kappa \, e^{ic_5 \eta_{55} \kappa \tau} \left[\delta\left(x^2 \right) + \theta\left(-\eta_{55} x^2 \right) \frac{\partial}{\partial x^2} J_0 \left(\kappa \left| x^2 \right|^{1/2} \right) \right].$$

Now performing the κ integration, the pre-Maxwell Green's function becomes

$$G(x, \tau) = -\frac{1}{2\pi} \delta(x^2) \delta(\tau) - \frac{c_5}{2\pi^2} \frac{\partial}{\partial x^2} \theta(-\eta_{55} g_{\alpha\beta} x^\alpha x^\beta) \frac{1}{\sqrt{-\eta_{55} g_{\alpha\beta} x^\alpha x^\beta}} \qquad (3.24)$$

so that both terms have units of distance^{-2} × time^{-1}. The first term contains the O(3,1) scalars x^2 and τ separately, and is called $G_{Maxwell}$. It has support at instantaneous τ and, as in Maxwell theory, along lightlike separations. The second term, called $G_{Correlation}$, has support determined by

$$-\eta_{55}\eta_{\alpha\beta}x^\alpha x^\beta = \begin{cases} -\left(x^2 + c_5^2\tau^2\right) = c^2t^2 - \mathbf{x}^2 - c_5^2\tau^2 > 0 \quad, \quad \eta_{55} = 1 \\ \left(x^2 - c_5^2\tau^2\right) = \mathbf{x}^2 - c^2t^2 - c_5^2\tau^2 > 0 \quad, \quad \eta_{55} = -1 \end{cases}$$

on timelike separations for $\eta^{55} = 1$ and spacelike separations for $\eta^{55} = -1$. Contributions from $G_{Correlation}$ are generally smaller than those of $G_{Maxwell}$ and drop off faster with distance from the source. To avoid singularities, particular care must be taken in handling the distribution functions. The derivative in $G_{Correlation}$ produces two singular terms

$$G_{Correlation}(x, \tau) = -\frac{c_5}{2\pi^2}\left(\frac{1}{2}\frac{\theta(-x^2 - c_5^2\tau^2)}{(-x^2 - c_5^2\tau^2)^{3/2}} - \frac{\delta\left(-x^2 - c_5^2\tau^2\right)}{\left(-x^2 - c_5^2\tau^2\right)^{1/2}}\right)$$

but these singularities cancel when first combined under integrals of the type (3.23) prior to applying the limits of integration. This order of operations expresses an aspect of the boundary conditions posed by Schwinger in deriving the Klein–Gordon Green's function.

3.6   THE MASS-ENERGY-MOMENTUM TENSOR

Under transformations $x^\alpha \to x'^\alpha = x^\alpha + \delta x^\alpha$ that leave the action invariant, a field undergoes

$$\phi(x) \to \phi'(x') = \phi(x) + \delta_0\phi(x) + \delta_x\phi(x) = \phi(x) + \delta_0\phi(x) + \delta x^\alpha \partial_\alpha\phi(x),$$

where

$$\delta_0\phi(x) = \phi'(x) - \phi(x)$$

is a variation in the form of the field at a fixed point x and

$$\delta_x\phi(x) = \delta x^\alpha \partial_\alpha\phi(x)$$

is a variation induced in the fixed form of the field by the variation of x. The action undergoes

$$\delta S_{em} = \int_{\Lambda'} d^4x'd\tau'L' - \int_{\Lambda} d^4x\,d\tau\,L,$$

where $\Lambda \to \Lambda'$ is the change of volume induced by the variation in x. Expanding the first term, this becomes

$$\delta S_{em} = -\int_{\Lambda} d^4x\,d\tau\,(\partial_\alpha L)\delta x^\alpha + \int_{\Lambda} d^4x\,d\tau\,\partial_\alpha\left(Lg^\alpha{}_\beta - \frac{\partial L}{\partial(\partial_a\phi)}\partial_\beta\phi(x)\right)\delta x^\beta$$

$$-\int_{\Lambda} d^4x\,d\tau\,\partial_\alpha\left(\frac{\partial L}{\partial(\partial_a\phi)}\delta x^\beta \delta_\beta\phi\right),$$

where we used the Euler–Lagrange equations

$$\frac{\partial L}{\partial \phi} - \partial_a \left(\frac{\partial L}{\partial (\partial_a \phi)} \right) = 0.$$

Since $\delta S = 0$ and the variations are arbitrary, we obtain Noether's theorem

$$\partial_\alpha \left(L g^{\alpha\beta} - \frac{\partial L}{\partial (\partial_a \phi)} \partial^\beta \phi (x) \right) = \partial_\alpha Q^{\alpha\beta} = 0$$

for the conserved current $Q^{\alpha\beta}$.

The electromagnetic Lagrangian can be written

$$L_{\text{em}} = \frac{e}{c^2} \, j^\alpha(x, \tau) a_\alpha(x, \tau) - \frac{1}{4c} f_\Phi^{\alpha\beta} (x, \tau) f_{\alpha\beta} (x, \tau),$$

where

$$f_\Phi^{\alpha\beta} (x, \tau) = \int \frac{ds}{\lambda} \, \Phi(\tau - s) \, f^{\alpha\beta} (x, s)$$

is the non-local convolved field. Under translations,

$$\delta x^\beta = \varepsilon^\beta \longrightarrow \delta a_\alpha = 0$$

and so the conserved current is

$$\tilde{\theta}_\Phi^{\alpha\beta} = \frac{\partial L}{\partial (\partial_\alpha a_\gamma)} \partial^\beta a_\gamma - L g^{\alpha\beta} = \frac{1}{c} g^{\alpha\beta} \left(\frac{1}{4} f_\Phi^{\delta\varepsilon} f_{\delta\varepsilon} - \frac{e}{c} j \cdot a \right) - \frac{1}{c} f_\Phi^{\alpha\gamma} \partial^\beta a_\gamma.$$

This current may be made symmetric in the indices by adding the total divergence

$$\Delta \theta_\Phi^{\alpha\beta} = \frac{1}{c} \partial_\gamma \left(f_\Phi^{\alpha\gamma} a^\beta \right) = \frac{e}{c^2} j^\alpha a^\beta + \frac{1}{c} f_\Phi^{\alpha\gamma} \partial_\gamma a^\beta,$$

where the second form follows from the inhomogeneous pre-Maxwell equation. Now, the symmetric current is

$$\theta_\Phi^{\alpha\beta} = \tilde{\theta}_\Phi^{\alpha\beta} + \Delta \theta_\Phi^{\alpha\beta} = \theta_{\Phi 0}^{\alpha\beta} + \frac{e}{c^2} \left[j^\alpha a^\beta - j \cdot a \, g^{\alpha\beta} \right],$$

where

$$\theta_{\Phi 0}^{\alpha\beta} = \frac{1}{c} \left[f_\Phi^{\alpha\gamma} f_\gamma^{\ \beta} + \frac{1}{4} f_\Phi^{\delta\varepsilon} f_{\delta\varepsilon} g^{\alpha\beta} \right]$$

is the source-free current. By explicit calculation, using the homogeneous pre-Maxwell equation, we find

$$\partial_\alpha T_\Phi^{\alpha\beta} = -\frac{e}{c^2} f^{\beta\alpha} j_\alpha,$$

where

$$T_\Phi^{\alpha\beta} = -\theta_{\Phi 0}^{\alpha\beta} = \frac{1}{c}\left[f_\Phi^{\alpha\gamma} f^\beta{}_\gamma - \frac{1}{4}g^{\alpha\beta} f_\Phi^{\delta\varepsilon} f_{\delta\varepsilon}\right]$$

is the conserved mass-energy-momentum tensor.

Writing the $\beta = 5$ component of the conservation law

$$\partial_\alpha T_\Phi^{\alpha 5} = -\frac{e}{c^2} f^{5\alpha} j_\alpha \qquad (3.25)$$

and using

$$j^\alpha(x,\tau) = c\dot{X}^\alpha(\tau)\delta^4\left(x - X(\tau)\right)$$

for the single particle current leads to

$$\partial_\alpha T_\Phi^{\alpha 5} = -\frac{e}{c} f^{5\alpha}(x,\tau)\dot{X}_\alpha(\tau)\delta^4\left(x - X(\tau)\right).$$

Integrating the LHS over spacetime leaves the τ-derivative

$$\int d^4x\, \partial_\alpha T_\Phi^{\alpha 5} = \int d^4x\, \partial_\mu T_\Phi^{\mu 5} + \frac{1}{c_5}\frac{d}{d\tau}\int d^4x\, T_\Phi^{55} = \frac{1}{c_5}\frac{d}{d\tau}\int d^4x\, T_\Phi^{55}$$

and integrating the RHS gives

$$-\frac{e}{c}\int d^4x\, f^{5\mu}(x,\tau)\dot{X}_\mu(\tau)\delta^4\left(x - X(\tau)\right) = -\frac{e}{c} f^{5\mu}(X(\tau),\tau)\dot{X}_\mu(\tau).$$

Recognizing this expression from the fifth Lorentz force equation

$$\frac{d}{d\tau}\left(-\frac{1}{2}M\dot{x}^2\right) = \eta_{55}\frac{ec_5}{c} f^{5\mu}\dot{x}_\mu$$

the RHS and LHS combine as

$$\frac{d}{d\tau}\left[\int d^4x\, T_\Phi^{55} + \eta_{55}\left(-\frac{1}{2}M\dot{x}^2\right)\right] = 0$$

demonstrating that the total mass of fields and events is conserved.

Since $M\dot{x}^2$ has units of energy ($\dot{x}^2 = -c^2$), we see that T_Φ^{55} has units of energy density (energy per 4D spacetime volume).

3.7 WORLDLINE CONCATENATION

We saw in (3.21) that the source current satisfies $\partial_\alpha j_\varphi^\alpha(x,\tau) = 0$, and so the vector part $j_\varphi^\mu(x,\tau)$ cannot be a divergenceless Maxwell current. However, Stueckelberg noticed that under the boundary condition

$$j_\varphi^5(x,\tau) \xrightarrow[\tau\to\pm\infty]{} 0$$

we have

$$\partial_\mu \int d\tau \, j_\varphi^\mu (x, \tau) + \frac{1}{c_5} \int d\tau \, \partial_\tau j_\varphi^5 (x, \tau) = \partial_\mu J^\mu (x) = 0,$$

where using (3.16) we confirm

$$J^\mu (x) = \int d\tau \, j_\varphi^\mu (x, \tau) = \int d\tau \int \frac{ds}{\lambda} \varphi (\tau - s) \, j^\alpha (x, s) = \int ds \, j^\mu (x, s)$$

in agreement with (3.9). Again, this integration, called concatenation [5], represents the sum at the spacetime point x of all events occurring over time τ. Saad and Horwitz [3] extended Stueckelberg's argument, showing that under the additional boundary condition

$$f^{5\mu}(x, \tau) \xrightarrow[\tau \to \pm\infty]{} 0$$

τ-integration of the pre-Maxwell equations leads to Maxwell's equations in the form

$$\left.\begin{array}{c} \partial_\beta f^{\alpha\beta} (x, \tau) = \dfrac{e}{c} j_\varphi^\alpha (x, \tau) \\[1em] \partial_{[\alpha} f_{\beta\gamma]} = 0 \\[1em] \partial_\alpha j^\alpha = 0 \end{array}\right\} \quad \xrightarrow{\displaystyle\int \frac{d\tau}{\lambda}} \quad \left\{\begin{array}{c} \partial_\nu F^{\mu\nu} (x) = \dfrac{e}{c} J^\mu (x) \\[1em] \partial_{[\mu} F_{\nu\rho]} = 0 \\[1em] \partial_\mu J^\mu (x) = 0, \end{array}\right.$$

where

$$F^{\alpha\nu}(x) = \int \frac{d\tau}{\lambda} \, f^{\alpha\nu}(x, \tau).$$

Under concatenation, $F^{\mu\nu}$ become Maxwell fields while $F^{5\mu}$ decouples from the Maxwell system. In addition, integrating the Green's function (3.24) for the pre-Maxwell wave equation

$$\int d\tau \, G_{Maxwell} = D(x) = -\frac{1}{2\pi}\delta(x^2) \qquad \int d\tau \, G_{Correlation} = 0 \qquad (3.26)$$

recovers the 4D Maxwell Green's function. When concatenating $G_{Correlation}$, the two singular terms arising from the derivative must once again be subtracted prior to applying the limits of integration.

As we have seen, SHP electrodynamics can be understood as a microscopic theory of events interacting at time τ. We saw in Section 3.6 that during these interactions, particles and pre-Maxwell fields may exchange mass under conservation of total mass-energy-momentum. As mentioned in Section 1.3, Feynman recognized that mass exchange of this type is also permitted, in principle, in QED. He interpreted integration over the evolution parameter, the final step in Equation (1.3) that describes the quantum Green's function for scalar particles, as extraction of asymptotic mass eigenstates from these complex interactions. In much the same way, we will see that concatenation—integration of the pre-Maxwell field equations over the evolution parameter τ—extracts from the microscopic event interactions the massless modes in Maxwell electrodynamics, expressing a certain equilibrium limit when mass exchange settles to zero. Thus, we will frequently compare the concatenated form of results in pre-Maxwell electrodynamics with the corresponding formulation in Maxwell theory, as a means of maintaining contact with established phenomenology.

3.8   PCT IN CLASSICAL SHP THEORY

We recall from Section 1.1 that Stueckelberg's initial motivation for consideration of τ-evolution was his desire to formulate a classical electrodynamics that includes antiparticles and describes pair processes through the dynamic evolution of a single type of evolution $x^\mu(\tau)$. And as we saw in Section 2.3, particles and antiparticles differ only in the direction of their t-evolution, specifically the sign of $\dot{x}^0(\tau)$, or equivalently the sign of the energy $E = p^0 c$. In quantum field theory the relationship of particles and antiparticles, characterized by a charge conjugation operation C, is significantly different. This operation is understood as a third discrete symmetry of the field equations, along with the improper Lorentz symmetries—time reversal T and space reversal P—that we expect to hold for a Lorentz covariant system. An antiparticle is obtained by acting with C and T to produce a particle with both the sign of the charge and the temporal ordering of its evolution reversed. Operator implementation of C generally requires a quantum formalism with complex wavefunctions, and the combined CT operation is anti-unitary.

We require electrodynamics to be symmetric under the improper Lorentz transformations T and P, and first find the transformations of the fields under these operations. We then consider a C operation, which in Wigner's original sense of "reversal of the direction of motion" [10], acts on the τ-evolution.

The Lorentz equations in explicit three-vector form are

$$M \frac{d^2 x^0}{d\tau^2} = \frac{e}{c} \left[\mathbf{e}(t, \mathbf{x}, \tau) \cdot \frac{d\mathbf{x}}{d\tau} - \eta_{55} \epsilon^0(t, \mathbf{x}, \tau) \right]$$

$$M \frac{d^2 \mathbf{x}}{d\tau^2} = \frac{e}{c} \left[\mathbf{e}(t, \mathbf{x}, \tau) \frac{dx^0}{d\tau} + \frac{d\mathbf{x}}{d\tau} \times \mathbf{b}(t, \mathbf{x}, \tau) - \eta_{55} \boldsymbol{\epsilon}(t, \mathbf{x}, \tau) \right]$$

and under space inversion P

$$x = \left(x^0, \mathbf{x} \right) \xrightarrow[P]{} x_P = \left(x_P^0, \mathbf{x}_P \right) = \left(x^0, -\mathbf{x} \right)$$

become

$$M \frac{d^2 x_P^0}{d\tau^2} = \frac{e}{c} \left[\mathbf{e}_P(t_P, \mathbf{x}_P, \tau) \cdot \frac{d\mathbf{x}_P}{d\tau} - \eta_{55} \epsilon_P^0(t_P, \mathbf{x}_P, \tau) \right]$$

$$M \frac{d^2 \mathbf{x}_P}{d\tau^2} = \frac{e}{c} \left[\mathbf{e}_P(t_P, \mathbf{x}_P, \tau) \frac{dx_P^0}{d\tau} + \frac{d\mathbf{x}_P}{d\tau} \times \mathbf{b}_P(t_P, \mathbf{x}_P, \tau) - \eta_{55} \boldsymbol{\epsilon}_P(t_P, \mathbf{x}_P, \tau) \right]$$

so that

$$M\frac{d^2x^0}{d\tau^2} = \frac{e}{c}\left[-\mathbf{e}_P\left(t_P, \mathbf{x}_P, \tau\right)\cdot\frac{d\mathbf{x}}{d\tau} - \eta_{55}\epsilon_P^0\left(t_P, \mathbf{x}_P, \tau\right)\right]$$

$$M\frac{d^2\mathbf{x}}{d\tau^2} = \frac{e}{c}\left[-\mathbf{e}_P\left(t_P, \mathbf{x}_P, \tau\right)\frac{dx^0}{d\tau} + \frac{d\mathbf{x}}{d\tau}\times\mathbf{b}_P\left(t_P, \mathbf{x}_P, \tau\right) - \eta_{55}\left(-\boldsymbol{\epsilon}\left(t_P, \mathbf{x}_P, \tau\right)\right)\right].$$

Invariance under P, understood as form invariance of the interaction, requires that

$$\mathbf{e}_P\left(t_P, \mathbf{x}_P, \tau\right) = -\mathbf{e}\left(t, \mathbf{x}, \tau\right) \qquad \mathbf{b}_P\left(t_P, \mathbf{x}_P, \tau\right) = \mathbf{b}\left(t, \mathbf{x}, \tau\right)$$

$$\epsilon_P^0\left(t_P, \mathbf{x}_P, \tau\right) = \epsilon^0\left(t, \mathbf{x}, \tau\right) \qquad \boldsymbol{\epsilon}_P\left(t_P, \mathbf{x}_P, \tau\right) = -\boldsymbol{\epsilon}\left(t, \mathbf{x}, \tau\right)$$

and as we would generally expect, the vectors $\mathbf{e}$ and $\boldsymbol{\epsilon}$ change sign, while the axial vector $\mathbf{b}$ and 0-component ϵ^0 are unchanged. Under time inversion T,

$$x = \left(x^0, \mathbf{x}\right) \xrightarrow{T} x_T = \left(x_T^0, \mathbf{x}_T\right) = \left(-x^0, \mathbf{x}\right);$$

we similarly write

$$M\frac{d^2x_T^0}{d\tau^2} = \frac{e}{c}\left[\mathbf{e}_T\left(t_T, \mathbf{x}_T, \tau\right)\cdot\frac{d\mathbf{x}_T}{d\tau} - \eta_{55}\epsilon_T^0\left(t_T, \mathbf{x}_T, \tau\right)\right]$$

$$M\frac{d^2\mathbf{x}_T}{d\tau^2} = \frac{e}{c}\left[\mathbf{e}_T\left(t_T, \mathbf{x}_T, \tau\right)\frac{dx_T^0}{d\tau} + \frac{d\mathbf{x}_T}{d\tau}\times\mathbf{b}_T\left(t_T, \mathbf{x}_T, \tau\right) - \eta_{55}\boldsymbol{\epsilon}_T\left(t_T, \mathbf{x}_T, \tau\right)\right]$$

so that

$$M\frac{d^2x^0}{d\tau^2} = \frac{e}{c}\left[-\mathbf{e}_T\left(t_T, \mathbf{x}_T, \tau\right)\cdot\frac{d\mathbf{x}}{d\tau} - \eta_{55}\left(-\epsilon_T^0\left(t_T, \mathbf{x}_T, \tau\right)\right)\right]$$

$$M\frac{d^2\mathbf{x}}{d\tau^2} = \frac{e}{c}\left[-\mathbf{e}_T\left(t_T, \mathbf{x}_T, \tau\right)\frac{dx^0}{d\tau} + \frac{d\mathbf{x}}{d\tau}\times\mathbf{b}_T\left(t_T, \mathbf{x}_T, \tau\right) - \eta_{55}\boldsymbol{\epsilon}_T\left(t_T, \mathbf{x}_T, \tau\right)\right].$$

Now form invariance requires

$$\mathbf{e}_T\left(t_T, \mathbf{x}_T, \tau\right) = -\mathbf{e}\left(t, \mathbf{x}, \tau\right) \qquad \mathbf{b}_T\left(t_T, \mathbf{x}_T, \tau\right) = \mathbf{b}\left(t, \mathbf{x}, \tau\right)$$

$$\epsilon_T^0\left(t_T, \mathbf{x}_T, \tau\right) = -\epsilon^0\left(t, \mathbf{x}, \tau\right) \qquad \boldsymbol{\epsilon}_T\left(t_T, \mathbf{x}_T, \tau\right) = \boldsymbol{\epsilon}\left(t, \mathbf{x}, \tau\right)$$

and here we notice that ϵ^0 and $\boldsymbol{\epsilon}$ transform as expected for components of a 4-vector, but the transformations of $\mathbf{e}$ and $\mathbf{b}$ are *opposite* to the behavior generally attributed to the electric and magnetic 3-vectors under time inversion. This can be attributed to our having respected the independence of $x^0\left(\tau\right)$ as a function of τ, not constrained by the mass-shell condition

$$\frac{dx^0}{d\tau} = +\frac{1}{\sqrt{1 - \left(d\mathbf{x}/dt\right)^2}}.$$

In general, all of the field components transform tensorially as components of the $f^{\mu\nu}$ and ϵ^{μ}.

From the transformation properties for the field strengths, we may deduce the transformation properties of the 5-vector potential components. First, we have

$$e^i_P = -e^i \qquad \Longrightarrow \qquad \partial^0 a^i_P - \left(-\partial^i\right) a^0_P = -\left(\partial^0 a^i - \partial^i a^0\right)$$

and so we conclude that

$$a^0_P = a^0 \qquad\qquad a^i_P = -a^i$$

which is consistent with

$$\mathbf{b}^i_P = \mathbf{b}^i = \varepsilon^{ijk}\partial_j a_k\ . \tag{3.27}$$

Similarly,

$$e^i_T = -e^i \qquad \Longrightarrow \qquad -\partial^0 a^i_T - \partial^i a^0_T = -\left(\partial^0 a^i - \partial^i a^0\right)$$

so we see that

$$a^0_T = -a^0 \qquad\qquad a^i_T = a^i$$

again consistent with (3.27). For the second vector field,

$$\epsilon^i_P = -\epsilon^i \qquad \Longrightarrow \qquad \partial^5 a^i_P - \left(-\partial^i\right) a^5_P = -\left(\partial^5 a^i - \partial^i a^5\right)$$

along with $a^i_P = -a^i$ leads to

$$a^5_P = a^5$$

which is consistent with

$$\epsilon^0_P = \epsilon^0 = \partial^5 a^0 - \partial^0 a^5 .$$

Similarly,

$$\epsilon^i_T = \epsilon^i \qquad \Longrightarrow \qquad \partial^5 a^i_T - \partial^i a^5_T = \partial^5 a^i - \partial^i a^5$$

along with $a^i_T = a^i$ leads to

$$a^5_T = a^5 .$$

Thus, the 4-vector and scalar components of the potential transform tensorially under space and time inversion.

The pre-Maxwell equations in 3-vector form, as given in (4.17) and (4.18),

$$\nabla \cdot \mathbf{e} - \frac{1}{c_5}\frac{\partial}{\partial\tau}\epsilon^0 = \frac{e}{c}j^0_\varphi = e\rho^0_\varphi \qquad\qquad\qquad \nabla\cdot\mathbf{b} = 0$$

$$\nabla\times\mathbf{b} - \frac{1}{c}\frac{\partial}{\partial t}\mathbf{e} - \frac{1}{c_5}\frac{\partial}{\partial\tau}\boldsymbol{\epsilon} = \frac{e}{c}\mathbf{j}_\varphi \qquad\qquad \nabla\times\mathbf{e} + \frac{1}{c}\frac{\partial}{\partial t}\mathbf{b} = 0$$

$$\nabla\cdot\boldsymbol{\epsilon} + \frac{1}{c}\frac{\partial}{\partial t}\epsilon^0 = \frac{e}{c}j^5_\varphi = \frac{ec_5}{c}\rho_\varphi \qquad\qquad \nabla\times\boldsymbol{\epsilon} - \eta^{55}\frac{1}{c_5}\frac{\partial}{\partial\tau}\mathbf{b} = 0$$

$$\nabla\epsilon^0 + \frac{1}{c}\frac{\partial}{\partial t}\boldsymbol{\epsilon} + \eta^{55}\frac{1}{c_5}\frac{\partial}{\partial\tau}\mathbf{e} = 0$$

are seen to be invariant under P and T using the transformations of the fields, under the choices

$$j_P^0(t_P, \mathbf{x}_P, \tau) = j^0(t, \mathbf{x}, \tau) \qquad j_T^0(t_T, \mathbf{x}_T, \tau) = -j^0(t, \mathbf{x}, \tau)$$

$$\mathbf{j}_P(t_P, \mathbf{x}_P, \tau) = -\mathbf{j}(t, \mathbf{x}, \tau) \qquad \mathbf{j}_T(t_T, \mathbf{x}_T, \tau) = \mathbf{j}(t, \mathbf{x}, \tau)$$

$$j_P^5(t_P, \mathbf{x}_P, \tau) = j^5(t, \mathbf{x}, \tau) \qquad j_T^5(t_T, \mathbf{x}_T, \tau) = j^5(t, \mathbf{x}, \tau),$$

where again the 4-vector and scalar components of the current transform tensorially under space and time inversion.

In order to discuss charge conjugation, we must make another short digression into quantum mechanics. As in Section 3.1, we may write the Stueckelberg–Schrodinger equation as

$$\left(i\partial_\tau + \frac{ec_5}{c}a_5\right)\psi(x,\tau) = \frac{1}{2M}\left(p^\mu - \frac{e}{c}a^\mu\right)\left(p_\mu - \frac{e}{c}a_\mu\right)\psi(x,\tau)$$
$$= -\frac{1}{2M}\left(\partial^\mu - \frac{ie}{c}a^\mu\right)\left(\partial_\mu - \frac{ie}{c}a_\mu\right)\psi(x,\tau)$$

and, taking the complex conjugate, observe that this system will be form invariant under a charge conjugation C that operates as

$$e \quad \underset{C}{\longrightarrow} \quad e_C = -e$$

$$\psi(x,\tau) \quad \underset{C}{\longrightarrow} \quad \psi_C(x,\tau) = \psi^*(x,-\tau)$$

$$\tau \quad \underset{C}{\longrightarrow} \quad \tau_C = -\tau$$

$$a^\mu(x,\tau) \quad \underset{C}{\longrightarrow} \quad a_C^\mu(x,\tau) = a^\mu(x,-\tau)$$

$$a^5(x,\tau) \quad \underset{C}{\longrightarrow} \quad a_C^5(x,\tau) = -a^5(x,-\tau)$$

if these transformations can be made consistent with the pre-Maxwell equations and Lorentz force. As we now show, this consistency can indeed be established. Leaving aside the quantum wavefunction and returning to classical mechanics, transformations of the potentials lead to field strength transformations

$$e^k = \partial^0 a^k - \partial^k a^0 \quad \underset{C}{\longrightarrow} \quad e^k$$

$$b^k = \varepsilon^{kij}\partial_i a_j \quad \underset{C}{\longrightarrow} \quad b^k$$

$$\epsilon^k = \eta_{55}\partial_\tau a^k - \partial^k a_5 \quad \underset{C}{\longrightarrow} \quad -\epsilon^k$$

$$\epsilon^0 = \eta_{55}\partial_\tau a^0 - \partial^0 a_5 \quad \underset{C}{\longrightarrow} \quad -\epsilon^0$$

so that this operation reverses the sign of tensor quantities carrying a scalar index. Under these transformations, the pre-Maxwell equations remain form invariant as long as

$$(j^0, \mathbf{j}, j^5) \quad \underset{C}{\longrightarrow} \quad (j^0, \mathbf{j}, j^5)_C = (j^0, \mathbf{j}, -j^5)$$

which is again a reversal of the scalar component. Similarly, the Lorentz force

$$M\frac{d^2x_C^0}{d\tau_C^2} = \frac{e}{c}\left[e_C \cdot \frac{d\mathbf{x}_C}{d\tau_C} - \eta_{55}\epsilon_C^0\right]$$

$$M\frac{d^2\mathbf{x}_C}{d\tau_C^2} = \frac{e}{c}\left[e_C\frac{dx_C^0}{d\tau_C} + \frac{d\mathbf{x}_C}{d\tau_C}\times\mathbf{b}_C - \eta_{55}\epsilon_C\right]$$

undergoes

$$M\frac{d^2x^0}{d\tau^2} = \frac{e}{c}\left[\mathbf{e}\cdot\left(-\frac{d\mathbf{x}}{d\tau}\right) - \eta_{55}\left(-\epsilon^0\right)\right]$$

$$M\frac{d^2\mathbf{x}}{d\tau^2} = \frac{e}{c}\left[\mathbf{e}\left(-\frac{dx^0}{d\tau}\right) + \left(-\frac{d\mathbf{x}}{d\tau}\right)\times\mathbf{b} - \eta_{55}\left(-\epsilon\right)\right]$$

becoming

$$M\frac{d^2x^0}{d\tau^2} = -\frac{e}{c}\left[\mathbf{e}\cdot\frac{d\mathbf{x}}{d\tau} - \eta_{55}\epsilon^0\right]$$

$$M\frac{d^2\mathbf{x}}{d\tau^2} = -\frac{e}{c}\left[\mathbf{e}\frac{dx^0}{d\tau} + \frac{d\mathbf{x}}{d\tau}\times\mathbf{b} - \eta_{55}\epsilon\right],$$

thus implementing classical charge conjugation. We see that current conservation

$$\partial_\mu j^\mu + \partial_\tau j^5 = 0 \quad \xrightarrow{C} \quad \partial_\mu j^\mu + (-\partial_\tau)\left(-j^5\right) = 0$$

is preserved, but since j^5 is interpreted as the number of events in a localized spacetime volume at a given τ, the meaning of $j_C^5 = -j^5$ must be examined carefully.

In standard relativistic mechanics, the continuity equation leads to a conserved charge through integration over volume in space as

$$\partial_\mu J^\mu = 0 \quad \longrightarrow \quad \frac{dQ}{d\tau} = \frac{d}{d\tau}\int d^3x\left(eJ^0\right) = -c\int d^3x\nabla\cdot(e\mathbf{J}) = 0$$

and since

$$J^0(x) = c\int d\tau\,\dot{X}^0(\tau)\,\delta^4(x - X(\tau))$$

cannot change sign in this approach, only the conjugation $e \to -e$ can account for charge reversal. But in SHP, charge conservation follows from

$$\partial_\alpha j^\alpha = 0 \quad \longrightarrow \quad \frac{dQ}{d\tau} = \frac{d}{d\tau}\int d^4x\left(ej^5\right) = -c_5\int d^4x\,e\,\partial_\mu j^\mu = 0,$$

where it is the event density

$$j^5(x) = c\dot{X}^5(\tau)\,\delta^4(x - X(\tau)) = cc_5\,\delta^4(x - X(\tau))$$

that cannot change sign. But the effective charge of an event interacting through the Lorentz force is associated with

$$ej^0(x) = ec\dot{X}^0(\tau)\,\delta^4(x - X(\tau))$$

which can change sign through $\dot{X}^0(\tau)$ according to Stueckelberg's prescription. Thus, the operation $e \to -e$ is not a required symmetry.

Following Stueckelberg, we disentangle the symmetries of the coordinate time t from those of the chronological parameter τ by making the following interpretations of the discrete reflections.

1. Space inversion covariance P implies certain symmetric relations between a given experiment and one performed in a spatially reversed configuration.

2. Time inversion covariance T implies certain symmetric relations between a given experiment and one performed in a t-reversed configuration, which is to say one in which advancement in t is replaced by retreat, and so a trajectory with $\dot{x}^0 > 0$ is replaced by a trajectory with $\dot{x}^0 < 0$. Thus, we expect symmetric behavior between pair annihilation processes and pair creation processes.

3. Charge conjugation covariance C implies certain symmetric relations between a given experiment and one in which the events are traced out in the reverse chronological order and carry opposite charge.

The operations P and T are improper Lorentz transformations and therefore must be symmetries of any (spinless Abelian) relativistic electrodynamics. But we do not regard the operation C defined here as connecting symmetrical dynamical evolutions. Rather, we associate the reversal of temporal order performed by C with the re-ordering of events performed by the observer in the laboratory, who interprets events as always evolving from earlier to later values of t. Thus, charge conjugation exchanges the viewpoint of the events under interaction with the viewpoint of the laboratory observer. The charge inversion (associated with the gauge symmetry) under this exchange reinforces the view of antiparticles in the laboratory, but does not influence the event dynamics.

3.9   BIBLIOGRAPHY

[1] Born, M. and Wolf, E. 1999. *Principles of Optics: Electromagnetic Theory of Propagation, Interference and Diffraction of Light*, Cambridge University Press, Cambridge. 25

[2] Jackson, J. D. and Okun, L. B. 2001. *Review of Modern Physics*, 73:663. 25

[3] Saad, D., Horwitz, L., and Arshansky, R. 1989. *Foundations of Physics*, 19:1125–1149. 27, 29, 37

[4] Jackson, J. 1975. *Classical Electrodynamics*, Wiley, New York. DOI: 10.1063/1.3057859. 29

[5] Arshansky, R., Horwitz, L., and Lavie, Y. 1983. *Foundations of Physics*, 13:1167. 29, 37

[6] Land, M. 1996. *Foundations of Physics*, 27:19. 29

[7] Land, M. 2003. *Foundations of Physics*, 33:1157. 29

[8] Land, M. 2017. *Entropy*, 19:234. http://dx.doi.org/10.3390/e19050234 32

[9] Schwinger, J. 1949. *Physical Review*, 75(4):651–679. https://link.aps.org/doi/10.1103/PhysRev.75.651 33

[10] Wigner, E. P. 1959. Group theory and its application to the quantum mechanics of atomic spectra, *Pure Applied Physics*, Academic Press, New York (translation from the German). https://cds.cern.ch/record/102713 38

PART III

Applications

CHAPTER 4

Problems in Electrostatics and Electrodynamics

4.1 THE COULOMB PROBLEM

Introductory treatments of electromagnetism quite naturally begin with the static Coulomb force between two point charges at rest. However, in the framework of Stueckelberg, Horwitz, and Piron, this seemingly simple configuration requires some clarification. A timelike event in its rest frame can be given with velocity

$$\dot{X}^2 = -c^2 \qquad \longrightarrow \qquad \dot{X} = (c, \mathbf{0})$$

so that this "static" event evolves uniformly in τ with coordinates

$$X(\tau) = (ct, \mathbf{X}) = (c(t_0 + \tau), \mathbf{X_0})$$

and the displacement $(ct_0, \mathbf{X_0})$ at $\tau = 0$ plays a role in interactions with other events.

Taking $\mathbf{X_0} = 0$, so that the event simply evolves along the t-axis in its rest frame, the associated event current is

$$j^\alpha(x, \tau) = c\dot{x}^\alpha \delta^4(x - X(\tau)) \longrightarrow \begin{cases} j^0(x, \tau) = c^2 \delta(ct - c(t_0 + \tau)) \delta^3(\mathbf{x}) \\ \mathbf{j}(x, \tau) = 0 \\ j^5(x, \tau) = cc_5 \delta(ct - c(t_0 + \tau)) \delta^3(\mathbf{x}) \end{cases}$$

with support restricted to the spatial origin—as in Maxwell theory—and to the time $t = t_0 + \tau$. The source for the pre-Maxwell field is the smoothed ensemble current

$$j_\varphi^\alpha(x, \tau) = \int ds\, \varphi(\tau - s)\, j^\alpha(x, s) \longrightarrow \begin{cases} j_\varphi^0 = c\varphi(t - (t_0 + \tau)) \delta^3(\mathbf{x}) \\ \mathbf{j}_\varphi = 0 \\ j_\varphi^5 = c_5\varphi(t - (t_0 + \tau)) \delta^3(\mathbf{x}) \end{cases} \tag{4.1}$$

which varies continuously in t, and as τ advances has its maximum at $t = t_0 + \tau$. The potential induced by this current may be found, as in (3.23), by integration with the Green's function (3.24), containing two terms, $G = G_{Maxwell} + G_{Correlation}$. We first treat the Maxwell term, which produces a potential with the expected $1/R$ dependence, multiplied by a time-dependent form factor found from φ. We then find the contribution from the correlation term which is scaled by the small factor c_5/c and drops off as $1/R^2$.

4.1.1 CONTRIBUTION TO POTENTIAL FROM $G_{Maxwell}$

The leading term in the potential is

$$a^0(x,\tau) = \frac{e}{2\pi} \int d^4x'd\tau' \, \delta\left((x-x')^2\right) \theta^{ret}\delta(\tau-\tau')\varphi\left(t'-(t_0+\tau')\right)\delta^3(x')$$

$$= \frac{e}{4\pi} \int cdt' \, \delta\left(c^2(t-t')^2 - x^2\right) \theta^{ret}\varphi\left(t'-(t_0+\tau)\right)$$

$$= \frac{e}{4\pi R}\varphi\left(\tau - \left(t - t_0 - \frac{R}{c}\right)\right)$$

$$\mathbf{a}(x,\tau) = 0$$

$$a^5(x,\tau) = \frac{c_5}{c}a^0(x,\tau),$$

where we insert $\theta^{ret} = \theta(x-x')$ to select retarded spacetime causality, and write $R = |x|$. As observed from a spacetime point $x = (ct, \mathbf{x})$, this field will grow as $\tau \to t - t_0 - R/c$ from below and then decrease. Since the time coordinate of the source is $t_{event} = t_0 + \tau$, the maximum occurs if the observer is located at time $t = t_{event} + R/c$, representing a delay equal to the signal transmission time at the speed of light. Put in a more familiar way, the time coordinate of the event detected at time t is $t_{event} = t_{retarded} = t - R/c$.

To study the "static" Coulomb problem, we consider a test event evolving uniformly at $x = (c\tau + ct_0^{test}, \mathbf{x})$, where $\mathbf{x}$ is constant. Inserting these coordinates and using (3.15), the potential experienced by the test event becomes

$$a^0(x,\tau) = \frac{e}{4\pi R}\varphi\left(-\Delta t_0 + \frac{R}{c}\right) = \frac{1}{2\xi}\frac{e}{4\pi R}e^{-|\Delta t_0 - R/c|/\xi\lambda} \tag{4.2}$$

$$a^5(x,\tau) = \frac{c_5}{c}a^0(x,\tau),$$

where $\Delta t_0 = t_0^{test} - t_0$ defines the mutual t-synchronization between the events. From (3.12) we may take $\xi \simeq 1/2$ for $c_5/c \ll 1$, and so recover the Coulomb potential

$$a^0(x,\tau) = \frac{e}{4\pi R}$$

in the particular case that $\Delta t_0 = R/c$. By contrast, if $\Delta t_0 = 0$, then a^0 takes the form of a Yukawa potential

$$a^0(x,\tau) = \frac{e}{4\pi R}e^{-2|R|/\lambda c} \tag{4.3}$$

suggesting a semi-classical interpretation in which the photons carrying the pre-Maxwell interaction have mass $m_\gamma c^2 \sim 2\hbar/\lambda$. Taking m_γ to be smaller than the experimental error on the mass of the photon $(10^{-18} eV/c^2)$ [1], we may estimate $\lambda > 10^4$ seconds. In this approximation λc will be larger than any practical distance in the problems we consider.

The field strength components found from the Yukawa-type potential with $\Delta t_0 = 0$ are

$$f^{k0}(x, \tau) = \partial^k \frac{e}{4\pi R} \frac{1}{2\xi} e^{-R/\xi \lambda c} \qquad f^{k5}(x, \tau) = \frac{c_5}{c} f^{k0}(x, \tau)$$

$$f^{ij}(x, \tau) = f^{50}(x, \tau) = 0$$

so that the test event will undergo Coulomb scattering

$$M \ddot{x}^k = \frac{e}{c} f^k{}_\nu \dot{x}^\nu - \eta_{55} \frac{e c_5}{c} f^{5k} = -\frac{e}{c} f^{k0} \left(\dot{x}^0 + \eta_{55} \frac{c_5^2}{c} \right)$$

according to the Lorentz force (3.4). Since the test event velocity is $\dot{x}(\tau) = (c, 0)$ this becomes

$$M \ddot{\mathbf{x}} = -\frac{e^2}{2\xi} \left(1 + \eta_{55} \left(\frac{c_5}{c} \right)^2 \right) \nabla \left(\frac{e^{-R/\xi \lambda c}}{4\pi R} \right) = -e^2 \frac{1 + \eta_{55} \left(\frac{c_5}{c} \right)^2}{1 + \left(\frac{c_5}{c} \right)^2} \nabla \left(\frac{e^{-R/\xi \lambda c}}{4\pi R} \right),$$

where we used (3.12) for ξ. Now suppose the source event were an antiparticle event evolving backward in time with $\dot{X}^0 = -c$. This would change the signs of $a^0(x, \tau)$ and $f^{k0}(x, \tau)$ but not the signs of $a^5(x, \tau)$ or $f^{k5}(x, \tau)$. We can thus write the Coulomb force for both cases as

$$\mathbf{F}^{(+/-)} = \mp e^2 \frac{1 \pm \eta_{55} \left(\frac{c_5}{c} \right)^2}{1 + \left(\frac{c_5}{c} \right)^2} \nabla \left(\frac{e^{-R/\xi \lambda c}}{4\pi R} \right),$$

where the upper sign is for a particle event and the lower sign is for an antiparticle event. Since $\eta_{55} = \pm 1$, this expression provides an experimental bound on c_5/c, given by

$$\frac{\sigma \left(e^- + e^+ \longrightarrow e^- + e^+ \right)}{\sigma \left(e^- + e^- \longrightarrow e^- + e^- \right)} = 1 \pm \text{experimental error} \simeq \left[\frac{1 \mp \eta_{55} \left(\frac{c_5}{c} \right)^2}{1 + \left(\frac{c_5}{c} \right)^2} \right]^2,$$

where σ is the total classical scattering cross-section at very low energy.

The action (3.13) recovers the usual first-order kinetic term $f^{\alpha\beta} f_{\alpha\beta}$ in the limit $\lambda \to 0$, in which case

$$\lim_{\lambda \to 0} \frac{1}{\xi \lambda} e^{-|\tau|/\xi \lambda} = \delta(\tau)$$

and the source of the pre-Maxwell field reduces to $j_\varphi^\alpha(x, \tau) \to j^\alpha(x, \tau)$. If we take λ very small but finite, then from (4.2) we have

$$a^0(x, \tau) \simeq -\frac{e}{4\pi R} \lambda \delta \left(\Delta t_0 - \frac{R}{c} \right)$$

for the potential experienced by the test event. Now the support of the potential is restricted to a lightline between the events, and for any synchronization $\Delta t_0 \neq R/c$ there will be no interaction. As we remarked in Section 3.3, a solution for Coulomb scattering can be found in this case [2], but the delta function potential leads to a discontinuous trajectory that is difficult to reconcile with classical phenomenology. This discontinuity is a primary motivation for introducing the interaction kernel. We mention in passing that this difficulty is not present in SHP quantum field theory because the definition of asymptotic states with sharp mass implies the loss of all information about the initial t-synchronization of the scattering particles.

The significance of the small λ limit appears in a number of places. As discussed in Section 3.4, λ characterizes the section of a worldline over which the event current is smoothed. In this sense, λ can be seen as the correlation length of a statistical process that assembles the current from an ensemble of events occurring along the trajectory. When λ is small, the interaction between an event trajectory and a test event is determined by a small number of points along the worldline, including only one point when $\lambda = 0$. Moreover, the mass spectrum $m_\gamma c^2 \sim 2\hbar/\lambda$ of the electromagnetic field associated with the Yukawa-like potential (4.3) becomes large.

By contrast, if λ is large, then the source $j_\varphi^\alpha(x, \tau)$ of the pre-Maxwell field is assembled from a large ensemble of events along the worldline, locally approximating the concatenation of the worldline performed in constructing the Maxwell current $J^\mu(x)$. In this case, the mass spectrum $m_\gamma c^2 \sim 2\hbar/\lambda$ of the electromagnetic field is small, approaching zero in the limit $\lambda \to \infty$.

From (3.16) and (4.1) the concatenated current is

$$J^0(x) = c \int \frac{d\tau}{\lambda} \, \varphi\left(t - (t_0 + \tau)\right) \delta^3(\mathbf{x}) = c\delta^3(\mathbf{x}) \qquad \mathbf{J}(x) = 0$$

describing a static Maxwell charge at the origin, and the concatenated potential is

$$A^0(x) = \frac{e}{4\pi R} \int \frac{d\tau}{\lambda} \, \varphi\left(\tau - \left(t - t_0 - \frac{R}{c}\right)\right) = \frac{e}{4\pi R} \qquad \mathbf{A}(x) = 0$$

describing the static Coulomb potential. As required, $J^\mu(x)$ and $A^\mu(x)$ are independent of t_0 and invariant under a shift of the event $x^\mu(\tau)$ along the time axis. The microscopic interaction between the events is thus seen to be sensitive to the t-synchronization Δt_0 of the interacting events, a parameter not accessible by the standard Coulomb law.

4.1.2 CONTRIBUTION TO POTENTIAL FROM $G_{Correlation}$

Up to this point, we have treated only the potential found from the leading term $G_{Maxwell}$ in the Green's function. To consider the potential found from $G_{Correlation}$ again we take as source the event $X = (ct_0 + c\tau, 0)$, but simplify the calculation by taking $t_0 = 0$ and approximate $\varphi(\tau' -$

$s) = \lambda\delta(\tau' - s)$ so that

$$a^0(x, \tau) = -\frac{e}{c}\int d^4x'd\tau' \, G_{Correlation}(x - x', \tau - \tau') \, c^2\lambda\delta\left(ct' - c\tau'\right)\delta^3(\mathbf{x}')$$

$$= -\lambda ec \int d\tau' \, G_{Correlation}\left((ct - c\tau', \mathbf{x}), \tau - \tau'\right).$$

We introduce the function $g(s)$ to express terms of the type

$$-\left((x - X(s))^2 + c_5^2(\tau - s)^2\right) = -\left(((ct, \mathbf{x}) - (cs, 0))^2 + c_5^2(\tau - s)^2\right) = c^2 g(s),$$

where

$$g(s) = (t - s)^2 - \frac{R^2}{c^2} - \frac{c_5^2}{c^2}(\tau - s)^2 = Cs^2 + Bs + A$$

and

$$\mu^2 = \frac{c_5^2}{c^2} \qquad C = \left(1 - \mu^2\right) \qquad B = -2\left(t - \mu^2\tau\right) \qquad A = t^2 - \frac{R^2}{c^2} - \mu^2\tau^2$$

so that the potential can be written as

$$a(x, \tau) = \frac{\lambda ec_5}{2\pi^2 c^3}(c, 0, c_5)\int ds \left[\frac{1}{2}\frac{\theta(g(s))}{g^{3/2}(s)} - \frac{\delta(g(s))}{g^{1/2}(s)}\right]\theta(t - s).$$

The zeros of $g(s)$ are found to be

$$s_\pm = \frac{-B \pm \sqrt{B^2 - 4AC}}{2C} = \frac{(t - \mu^2\tau) \pm \sqrt{\frac{R^2}{c^2}(1 - \mu^2) + \mu^2(t - \tau)^2}}{(1 - \mu^2)} \tag{4.4}$$

and since we assume $\mu^2 < 1$ there will be roots for any values of t and R. In addition, the condition $\theta^{ret} = \theta(t - s)$ requires $t > s$.

Attempting to set $t < s_-$ leads to

$$t < \frac{(t - \mu^2\tau) - \sqrt{\frac{R^2}{c^2}(1 - \mu^2) + \mu^2(t - \tau)^2}}{(1 - \mu^2)} \qquad \Rightarrow \qquad -\mu^2(t - \tau)^2 > \frac{R^2}{c^2}$$

and so $t \geq s_-$ is a condition of integration for the θ term. Similarly, attempting to set $t > s_+$ leads to

$$t > \frac{(t - \mu^2\tau) + \sqrt{\frac{R^2}{c^2}(1 - \mu^2) + \mu^2(t - \tau)^2}}{(1 - \mu^2)} \qquad \Rightarrow \qquad -\mu^2(t - \tau) > \frac{R^2}{c^2}$$

leading to the condition

$$\theta(g(s))\,\theta(t - s) \neq 0 \qquad \Rightarrow \qquad s_- \leq s \leq t \leq s_+$$

from which

$$a\left(x,\tau\right)=\frac{\lambda e c_5}{2\pi^2 c^3}\left(1,0,\frac{c_5}{c}\right)\left(\frac{1}{2}\int_{-\infty}^{s_-}ds\frac{1}{g^{3/2}\left(s\right)}-\int_{-\infty}^{\infty}ds\frac{\delta\left(g\left(s\right)\right)}{g^{1/2}\left(s\right)}\theta\left(t-s\right)\right).$$

Using the well-known form [3]

$$\int\frac{dx}{\left(Cx^2+Bx+A\right)^{3/2}}=\frac{2\left(2Cs+B\right)}{q(Cx^2+Bx+A)^{1/2}},$$

where

$$q=4AC-B^2$$

we notice from (4.4) that

$$s_-=\frac{-B-\sqrt{B^2-4AC}}{2C}=\frac{-B-\sqrt{-q}}{2C}\qquad\Rightarrow\qquad-\sqrt{-q}=2Cs_-+B$$

and so

$$\frac{1}{2}\int_{-\infty}^{s_-}ds\frac{1}{g^{3/2}\left(s\right)}=\frac{2Cs_-+B}{qg^{1/2}\left(s_-\right)}-\frac{2Cs+B}{qg^{1/2}\left(s\right)}\bigg|_{-\infty}$$

$$=\frac{-\sqrt{-q}}{qg^{1/2}\left(s_-\right)}+\frac{2\sqrt{C}}{\left(2Cs_-+B\right)^2}$$

$$=\frac{1}{\sqrt{-q}g^{1/2}\left(s_-\right)}+\frac{1}{2}\frac{\sqrt{1-\mu^2}}{\frac{R^2}{c^2}\left(1-\mu^2\right)+\mu^2\left(t-\tau\right)^2}.\qquad(4.5)$$

The second term is

$$\int_{-\infty}^{\infty}ds\frac{\delta\left(g\left(s\right)\right)}{g^{1/2}\left(s\right)}\theta\left(t-s\right)$$

and using the identity

$$\int ds\,f\left(s\right)\delta\left(g\left(s\right)\right)=\frac{f\left(s^-\right)}{\left|g'\left(s^-\right)\right|}\bigg|_{s^-=g^{-1}(0)}$$

we can evaluate

$$\int_{-\infty}^{\infty}ds\frac{\delta\left(g\left(s\right)\right)}{g^{1/2}\left(s\right)}\theta\left(t-s\right)=\frac{\theta\left(t-s_-\right)}{\left|g'\left(s_-\right)\right|g^{1/2}\left(s_-\right)}=\frac{1}{\left|g'\left(s_-\right)\right|g^{1/2}\left(s_-\right)}.$$

Since

$$g'\left(s_-\right)=\left(Cs_-^2+Bs_-+A\right)'=2Cs_-+B=-\sqrt{-q}$$

we see that this term cancels the singularity in the first term, leaving

$$\frac{1}{2}\int_{-\infty}^{s_-}ds\frac{1}{g^{3/2}\left(s\right)}-\int_{-\infty}^{\infty}ds\frac{\delta\left(g\left(s\right)\right)}{g^{1/2}\left(s\right)}\theta\left(t-s\right)=\frac{1}{2}\frac{\sqrt{1-\mu^2}}{\frac{R^2}{c^2}\left(1-\mu^2\right)+\mu^2\left(t-\tau\right)^2}$$

and

$$a\left(x,\tau\right) = \frac{\lambda e}{4\pi^2}\left(c,\mathbf{0},c_5\right)\frac{c_5}{c}\frac{\sqrt{1-\dfrac{c_5}{c}}}{R^2\left(1-\dfrac{c_5}{c}\right)+\dfrac{c_5}{c}c^2\left(t-\tau\right)^2}.$$

We notice that the potential has units of $\lambda c/\text{distance}^2 = 1/\text{distance}$, as does the potential associated with $G_{Maxwell}$. This contribution to the potential is smaller by a factor of c_5/c than the Yukawa potential found in (4.3), and drops off faster with distance, as $1/R^2$ compared to $1/R$. This term may be neglected when the contribution from $G_{Maxwell}$ is significant, but as we will see in Section 4.7.1, it may lead to qualitatively important phenomena when the leading term vanishes.

4.2   LIÉNARD–WIECHART POTENTIAL AND FIELD STRENGTH

We now consider an arbitrary event $X^\alpha\left(\tau\right)$ for which the smoothed current is

$$j_\varphi^\alpha\left(x,\tau\right) = c\int ds\,\varphi\left(\tau-s\right)\dot{X}^\alpha\left(s\right)\delta^4\left[x-X\left(s\right)\right]$$

and the Liénard–Wiechert potential found from $G_{Maxwell}$ is

$$\begin{aligned}a^\alpha\left(x,\tau\right) &= \frac{e}{2\pi c}\int d^4x'd\tau'\delta\left(\left(x-x'\right)^2\right)\theta^{ret}\delta\left(\tau-\tau'\right)j_\varphi^\alpha\left(x',\tau'\right)\\ &= \frac{e}{2\pi}\int ds\,\varphi\left(\tau-s\right)\dot{X}^\alpha\left(s\right)\delta\left(\left(x-X\left(s\right)\right)^2\right)\theta^{ret},\end{aligned} \tag{4.6}$$

where θ^{ret} imposes retarded x^0 causality. Writing the line of observation as

$$z^\mu = x^\mu - X^\mu(s) \qquad \longrightarrow \qquad z^2 = \left[x-X\left(s\right)\right]^2$$

and using the identity

$$\int ds\,f\left(s\right)\delta\left[g\left(s\right)\right] = \frac{f\left(\tau_R\right)}{\left|g'\left(\tau_R\right)\right|}\quad,\quad \tau_R = g^{-1}\left(0\right) \tag{4.7}$$

we obtain

$$a^\alpha\left(x,\tau\right) = \frac{e}{4\pi}\varphi\left(\tau-\tau_R\right)\frac{\dot{X}^\alpha\left(\tau_R\right)}{\left|\left(x^\mu - X^\mu\left(\tau_R\right)\right)\dot{X}_\mu\left(\tau_R\right)\right|}, \tag{4.8}$$

where the retarded time τ_R satisfies

$$z^2 = \left[x-X(\tau_R)\right]^2 = 0 \qquad \theta\left(x-X\left(\tau_R\right)\right) = 1.$$

Introducing the notation for velocity

$$u^\mu = \dot{X}^\mu (\tau) \qquad \beta^\mu = \frac{\dot{X}^\mu}{c} \qquad u^5 = \dot{X}^5 = c_5$$

and the scalar length

$$R = \frac{1}{2c}\frac{d}{d\tau_R}[x - X(\tau_R)]^2 = -\frac{z^\mu u_\mu}{c} = |z \cdot \beta| \tag{4.9}$$

the potential becomes

$$a^\mu (x, \tau) = \frac{e}{4\pi R}\varphi(\tau - \tau_R)\beta^\mu \qquad a^5 (x, \tau) = \frac{e}{4\pi R}\varphi(\tau - \tau_R)\frac{c_5}{c}, \tag{4.10}$$

where R is nonnegative because u^μ is timelike and z^μ is lightlike. Thus, $a^\mu (x, \tau)$ takes the form of the usual Liénard–Wiechert potential from Maxwell theory multiplied by the factor $\varphi(\tau - \tau_R)$ which separates out the τ-dependence of the fields.

To calculate the field strengths, we need derivatives of the Liénard–Wiechert potential. Since

$$\frac{d}{d\tau_R}\varphi(\tau - \tau_R) = -\frac{1}{2\xi}\frac{d}{d\tau}e^{-|\tau - \tau_R|/\xi\lambda} = -\frac{\varepsilon(\tau - \tau_R)}{\xi\lambda}\varphi(\tau - \tau_R),$$

where $\varepsilon(\tau) = \text{signum}(\tau)$, we obtain the τ-derivative

$$\frac{1}{c_5}\partial_\tau a_\mu (x, \tau) = \frac{1}{c_5}\frac{e}{4\pi|u \cdot z|}\dot{\varphi}(\tau - \tau_R)u_\mu = -\frac{e}{4\pi c_5}\frac{\varphi(\tau - \tau_R)}{|u \cdot z|}\frac{\varepsilon(\tau - \tau_R)}{\xi\lambda}u_\mu$$

directly from (4.10).

The spacetime derivative is most conveniently found by applying the identity (4.7) to expression (4.6)

$$\partial^\mu a^\beta (x, \tau) = \frac{e}{2\pi}\int ds\,\varphi(\tau - s)\dot{X}^\alpha (s)\,\theta^{ret}\,\partial^\mu \delta\left((x - X(s))^2\right)$$

$$= \frac{e}{2\pi}\int ds\,\varphi(\tau - s)\dot{X}^\beta (s)\,\theta^{ret}\,\delta'\left[(x - X(s))^2\right][2(x^\mu - X^\mu (s))]$$

$$= -\frac{e}{2\pi}\int ds\,\varphi(\tau - s)\frac{\dot{X}^\beta (s)[x^\mu - X^\mu (s)]}{\dot{X}(s) \cdot (x - X(s))}\theta^{ret}\frac{d}{ds}\delta\left[(x - X(s))^2\right]$$

and integrating by parts to obtain

$$\partial^\mu a^\beta (x, \tau) = \frac{e}{2\pi}\int ds\frac{d}{ds}\left[\varphi(\tau - s)\frac{\dot{X}^\beta (s)[x^\mu - X^\mu (s)]}{\dot{X}(s) \cdot (x - X(s))}\right]\theta^{ret}\delta\left[(x - X(s))^2\right]$$

$$= \frac{e}{4\pi}\frac{1}{|u \cdot z|}\frac{d}{ds}\left[\varphi(\tau - s)\frac{z^\mu (s)u^\beta (s)}{u \cdot z}\right]_{s=\tau_R}.$$

Using

$$\dot{z}^\mu = -u^\mu \qquad \dot{R} = -\frac{d}{d\tau}\frac{z \cdot u}{c} = c\beta^2 - z \cdot \dot{\beta}$$

we find the field strengths as

$$f^{\mu\nu}(x,\tau) = \frac{e}{4\pi}\frac{\varphi(\tau - \tau_R)}{R}\left\{\frac{(z^\mu \beta^\nu - z^\nu \beta^\mu)\beta^2}{R^2} - \frac{\varepsilon(\tau - \tau_R)}{\xi\lambda c}\frac{z^\mu \beta^\nu - z^\nu \beta^\mu}{R}\right.$$
$$\left. - \frac{\left(z^\mu \dot{\beta}^\nu - z^\nu \dot{\beta}^\mu\right)R + (z^\mu \beta^\nu - z^\nu \beta^\mu)\left(\dot{\beta} \cdot z\right)}{cR^2}\right\} \tag{4.11}$$

$$f^{5\mu}(x,\tau) = c_5\frac{e}{4\pi}\frac{\varphi(\tau - \tau_R)}{cR}\left\{-\frac{z^\mu \beta^2 + \beta^\mu R}{R^2} - \frac{\varepsilon(\tau - \tau_R)}{\xi\lambda c}\frac{z^\mu + \beta^\mu Rc^2/c_5^2}{R}\right.$$
$$\left. + \frac{z^\mu\left(\dot{\beta} \cdot z\right)}{cR^2}\right\}. \tag{4.12}$$

It is convenient to express the fields as elements of a Clifford algebra [4] with basis vectors

$$\mathbf{e}_\alpha \cdot \mathbf{e}_\beta = \eta_{\alpha\beta} \qquad \mathbf{e}_\alpha \wedge \mathbf{e}_\beta = \mathbf{e}_\alpha \otimes \mathbf{e}_\beta - \mathbf{e}_\beta \otimes \mathbf{e}_\alpha \tag{4.13}$$

and Clifford product

$$\mathbf{e}_\alpha \mathbf{e}_\beta = \mathbf{e}_\alpha \cdot \mathbf{e}_\beta + \mathbf{e}_\alpha \wedge \mathbf{e}_\beta.$$

Separating spacetime and scalar quantities as

$$X(\tau) = X^\mu(\tau)\mathbf{e}_\mu \qquad X^5 = c_5\tau$$
$$d = \partial_\mu e^\mu \qquad \partial_5 = \frac{1}{c_5}\partial_\tau$$

and writing $\epsilon^\mu = f^{5\mu}$, the field strength tensors

$$f = \frac{1}{2} f^{\mu\nu}\, \mathbf{e}_\mu \wedge \mathbf{e}_\nu \qquad f^5 = f^{5\mu}\, \mathbf{e}_5 \wedge \mathbf{e}_\mu = \mathbf{e}_5 \wedge \epsilon$$

are Clifford bivectors, (3.5) takes the form

$$f = d \wedge a \qquad \epsilon = -\partial_5 a - da^5.$$

In this notation, the pre-Maxwell equations (3.20) are

$$-d \cdot f - \partial_5 \epsilon = \frac{e}{c} j_\varphi \qquad d \cdot \epsilon = \frac{e}{c} j_\varphi^5$$
$$d \wedge f = 0 \qquad d \wedge \epsilon + \partial_5 f = 0, \tag{4.14}$$

where we may evaluate $d \cdot f$ and similar terms using the Clifford identity

$$a \cdot (b \wedge c) = (a \cdot b) c - (a \cdot c) b.$$

Defining the dimensionless quantities associated with acceleration $\dot{\beta} = \ddot{X}/c$,

$$Q = -\frac{\dot{\beta} \cdot z}{c} \qquad W = \frac{\dot{\beta} R - \beta c Q}{c} = -\frac{\dot{\beta} (\beta \cdot z) - \beta \left(\dot{\beta} \cdot z\right)}{c}$$

the field strengths become

$$f(x, \tau) = \frac{e}{4\pi} \varphi (\tau - \tau_R) \frac{z}{R^3} \wedge \left\{ \beta \left(\beta^2 - \frac{\varepsilon (\tau - \tau_R)}{\xi \lambda c} R \right) - W \right\}$$

$$\epsilon(x, \tau) = \frac{c_5}{c} \frac{e}{4\pi} \frac{\varphi (\tau - \tau_R)}{R} \left\{ -\frac{z\beta^2 + \beta R}{R^2} - \frac{\varepsilon (\tau - \tau_R)}{\xi \lambda c} \left(\frac{z}{R} + \beta \frac{c^2}{c_5^2} \right) - \frac{zQ}{R^2} \right\}$$

in which the factors $\varphi (\tau - \tau_R)$ and $\varepsilon (\tau - \tau_R)$ contain the τ-dependence. Since

$$\int \frac{d\tau}{\lambda} \varphi (\tau) = 1 \qquad -\int \frac{d\tau}{\lambda} \varphi (\tau) \frac{\varepsilon (\tau)}{\xi \lambda} = \int \frac{d\tau}{\lambda} \varphi' (\tau) = 0$$

the concatenated fields are found by replacing $\varphi (\tau - \tau_R) \to 1$ and $\varepsilon (\tau - \tau_R) \to 0$, in agreement with the standard Maxwell result. We mention again that these field strengths were obtained using only the leading term $G_{Maxwell}$ in the Green's function, and neglect the smaller contributions from $G_{Correlation}$. Although the neglected terms vanish under concatenation, they may dominate the dynamics when the leading contribution is zero. In particular, while $G_{Maxwell}$ has support on the lightcone, $G_{Correlation}$ has timelike or spacelike support (depending on the choice of η_{55}) and so becomes significant in self-interactions.

Taking $\lambda c \gg R$ and neglecting mass transfer, so that $\beta^2 = u^2/c^2 = -1$, we may approximate

$$f(x, \tau) = -\frac{e}{4\pi} \varphi (\tau - \tau_R) \frac{z}{R^3} \wedge (\beta + W) \tag{4.15}$$

$$\epsilon(x, \tau) = \frac{c_5}{c} \frac{e}{4\pi} \varphi (\tau - \tau_R) \frac{z (1 - Q) - \beta R}{R^3}$$

and split the field strengths into the short-range retarded fields

$$f^{ret} = -\frac{e}{4\pi} \varphi (\tau - \tau_R) \frac{z \wedge \beta}{R^3} \qquad \epsilon^{ret} = \frac{c_5}{c} \frac{e}{4\pi} \varphi (\tau - \tau_R) \frac{z - \beta R}{R^3}$$

that drop off as $1/R^2$, and the radiation fields

$$f^{rad} = -\frac{e}{4\pi} \varphi (\tau - \tau_R) \frac{z \wedge W}{R^3} \qquad \epsilon^{rad} = -\frac{c_5}{c} \frac{e}{4\pi} \varphi (\tau - \tau_R) \frac{zQ}{R^3}$$

associated with acceleration that drop off as $1/R$.

As elements of a Clifford algebra, the field strengths admit geometrical interpretation. The factor $z \wedge \beta$ in f^{ret} represents the plane spanned by the velocity β and the line of observation z. Similarly, we recognize

$$z - \beta R = -\beta^2 z + \beta (z \cdot \beta) = -(z \wedge \beta) \cdot \beta$$

representing the projection of β onto the $z - \beta$ plane, and so we have

$$f^{ret} = -\frac{e}{4\pi} \varphi (\tau - \tau_R) \frac{z \wedge \beta}{R^3} \qquad \epsilon^{ret} = \frac{c_5}{c} f^{ret} \cdot \beta$$

for the retarded fields. Similarly, using

$$a \cdot (b \wedge c \wedge d) = (a \cdot b) c \wedge d - (a \cdot c) b \wedge d + (a \cdot d) b \wedge c$$

and $z^2 = 0$, we see that

$$z \wedge W = \left(z \wedge \beta \wedge \dot{\beta} \right) \cdot z$$

in f^{rad} represent the projection of z onto the volume spanned by z, β, and $\dot{\beta}$. Similarly, ϵ^{ret} is proportional to $zQ = (\dot{\beta} \cdot z) z / c$, the projection of z onto the acceleration $\dot{\beta}$.

4.3 ELECTROSTATICS

The covariant equivalent of a spatially static charge is a uniformly evolving event

$$X (\tau) = u\tau = \left(u^0 \tau, \mathbf{u}\tau \right)$$

with constant timelike velocity $\dot{X} = u = \beta c$, which in its rest frame simply advances along the time axis as $t = \beta^0 \tau$. As a result, and given the geometrical interpretation of the Clifford forms, the field strengths are essentially kinematical in structure.

Writing the timelike velocity β in terms of the unit vector $\hat{\beta}$

$$\beta^2 < 0 \qquad \beta = |\beta| \hat{\beta} \qquad \hat{\beta}^2 = -1 \qquad \beta^2 = -|\beta|^2$$

the observation line z can be separated into components

$$z_\| = -\hat{\beta} \left(\hat{\beta} \cdot z \right) \qquad z_\perp = z + \hat{\beta} \left(\hat{\beta} \cdot z \right)$$

which satisfy

$$z_\|^2 = \hat{\beta}^2 \left(\hat{\beta} \cdot z \right)^2 = -\left(\hat{\beta} \cdot z \right)^2$$

$$z_\perp^2 = z^2 + 2 \left(\hat{\beta} \cdot z \right)^2 - \left(\hat{\beta} \cdot z \right)^2 = \left(\hat{\beta} \cdot z \right)^2 = -z_\|^2 \qquad (4.16)$$

$$(\beta \cdot z)^2 = |\beta|^2 \left(\hat{\beta} \cdot z \right)^2 = -|\beta|^2 z_\|^2 .$$

The condition of retarded causality

$$z^2 = c^2 \tau_R^2 \beta^2 - 2c\tau_R \beta \cdot x + x^2 = 0$$

relates the field to the location of the event along the backward lightcone of the observation point. This implicit choice of τ_R and its gradient

$$0 = d(z^2) = 2\left(c^2 \tau_R d\tau_R \beta^2 - c\tau_R \beta - cd\tau_R \beta \cdot x + x\right) = 2\left[cRd\tau_R + z\right]$$

lead to the following expressions:

$$d\tau_R = -\frac{z}{cR} \qquad (\beta \cdot d)\,\tau_R = \frac{\beta \cdot z}{\beta \cdot z} = 1 \qquad (z \cdot d)\,\tau_R = -\frac{z^2}{cR} = 0$$

$$d\,(\beta \cdot z) = d\left(\beta \cdot x - \beta^2 \tau_R\right) = \frac{(\beta \cdot z)\,\beta - \beta^2 z}{\beta \cdot z} = -\left|\beta^2\right|\frac{z_\perp}{cR}$$

$$d\frac{1}{R^n} = (-1)^n \frac{-n\left[(\beta \cdot z)\,\beta - \beta^2 z\right]}{(\beta \cdot z)^{n+2}} = \frac{-n\left|\beta^2\right|z_\perp}{R^{n+2}}$$

$$d \cdot z = d \cdot (x - c\beta\tau_R) = d \cdot x - c\beta \cdot d\tau_R = 3$$

$$d \wedge z = d \wedge (x - c\beta\tau_R) = cd\tau_R \wedge \beta = -\frac{\beta \wedge z}{R}$$

$$d \wedge \hat{z} = d \wedge \frac{z}{|z|} = -\frac{1}{|z|}\frac{\beta \wedge z}{R} - \hat{z} \wedge \frac{z}{|z|^2} = -\frac{\beta \wedge \hat{z}}{R}.$$

Using these expressions, the pre-Maxwell equations (4.14) can be easily verified for the case of a uniform velocity event [5]. For example, recalling $\varphi' = -\varphi\varepsilon/\xi\lambda$, the exterior derivative of f is

$$d \wedge f = \frac{e}{4\pi}d \wedge \left(\varphi\,(\tau - \tau_R)\frac{z \wedge \beta}{R^3}\beta^2 + \varphi'\,(\tau - \tau_R)\frac{z \wedge \beta}{cR^2}\right)$$

which produces terms of the type:

$$d\varphi^{(n)} \wedge (z \wedge \beta) = -\varphi^{(n+1)}\frac{z}{cR} \wedge (z \wedge u) = 0$$

$$d \wedge (z \wedge \beta) = (d \wedge z) \wedge \beta = -\frac{\beta \wedge z}{R} \wedge \beta = 0$$

$$\left[d\frac{1}{R^n}\right] \wedge (z \wedge u) = \left[\frac{-n\left|\beta^2\right|z_\perp}{R^{n+2}}\right] \wedge (z_\perp \wedge u) = 0$$

and thus we recover

$$d \wedge f = 0$$

from kinematics.

It is convenient to write the field strengths in 3-vector and scalar form

$$(\mathbf{e})^i = f^{0i} \qquad (\mathbf{b})_i = \varepsilon_{ijk} f^{jk} \qquad (\boldsymbol{\epsilon})^i = f^{5i} \qquad \epsilon^0 = f^{50}$$

for which the field equations split into four generalizations of the 3-vector Maxwell equations

$$\nabla \cdot \mathbf{e} - \frac{1}{c_5}\frac{\partial}{\partial \tau}\epsilon^0 = \frac{e}{c}j_\varphi^0 = e\rho_\varphi^0 \qquad\qquad \nabla \cdot \mathbf{b} = 0$$

$$\nabla \times \mathbf{b} - \frac{1}{c}\frac{\partial}{\partial t}\mathbf{e} - \frac{1}{c_5}\frac{\partial}{\partial \tau}\boldsymbol{\epsilon} = \frac{e}{c}\mathbf{j}_\varphi \qquad\qquad \nabla \times \mathbf{e} + \frac{1}{c}\frac{\partial}{\partial t}\mathbf{b} = 0$$

(4.17)

and three new equations for the fields $\boldsymbol{\epsilon}$ and ϵ^0

$$\nabla \cdot \boldsymbol{\epsilon} + \frac{1}{c}\frac{\partial}{\partial t}\epsilon^0 = \frac{e}{c}j_\varphi^5 = \frac{ec_5}{c}\rho_\varphi \qquad\qquad \nabla \times \boldsymbol{\epsilon} - \eta^{55}\frac{1}{c_5}\frac{\partial}{\partial \tau}\mathbf{b} = 0$$

$$\nabla\epsilon^0 + \frac{1}{c}\frac{\partial}{\partial t}\boldsymbol{\epsilon} + \eta^{55}\frac{1}{c_5}\frac{\partial}{\partial \tau}\mathbf{e} = 0.$$

(4.18)

Writing $d = \mathbf{e}_0\partial_0 + \nabla$ and $f = \mathbf{e}_0 \wedge \mathbf{e} + \frac{1}{2}f^{jk}\mathbf{e}_j \wedge \mathbf{e}_k$ we find that

$$d \wedge f = 0 \qquad \longrightarrow \qquad \begin{cases} \nabla \cdot \mathbf{b} = 0 \\[2mm] \nabla \times \mathbf{e} + \dfrac{1}{c}\dfrac{\partial}{\partial t}\mathbf{b} = 0 \end{cases}$$

expressing the absence of electromagnetic monopoles.

In the rest frame of a charged event, we may set $\dot{t} = 1 \rightarrow \beta = \mathbf{e}_0$, so for an observation point $x = (ct, \mathbf{x})$

$$z^2 = 0 \qquad \longrightarrow \qquad \begin{cases} \tau_R = t - \dfrac{|\mathbf{x}|}{c} \\[3mm] R = -\mathbf{e}_0 \cdot (x - c\tau_R\mathbf{e}_0) = |\mathbf{x}| \\[3mm] z = (c(t - \tau_R), \mathbf{x}) = R(\mathbf{e}_0 + \hat{\mathbf{x}}) \end{cases}$$

and the field strengths reduce to

$$f(x, \tau) = \frac{e}{4\pi}\varphi(\tau - \tau_R)\frac{\mathbf{e}_0 \wedge \hat{\mathbf{x}}}{R^2}\left(1 + \frac{\varepsilon(\tau - \tau_R)}{\xi\lambda c}R\right) = \mathbf{e}_0 \wedge \mathbf{e}(x, \tau)$$

$$\epsilon(x, \tau) = \frac{c_5}{c}\frac{e}{4\pi}\varphi(\tau - \tau_R)\left\{\frac{\hat{\mathbf{x}}}{R^2} + \frac{\varepsilon(\tau - \tau_R)}{\xi\lambda cR}\left[\mathbf{e}_0\left(1 + \frac{c^2}{c_5^2}\right) + \hat{\mathbf{x}}\right]\right\}.$$

We thus find that the magnetic field **b** is zero, while

$$\mathbf{e} = \frac{e}{4\pi}\left(\frac{\varphi\,(\tau - \tau_R)}{R^2} - \frac{\varphi'\,(\tau - \tau_R)}{R}\right)\hat{\mathbf{x}} \qquad \boldsymbol{\epsilon} = \frac{c_5}{c}\mathbf{e} \tag{4.19}$$

and

$$\epsilon^0(x,\tau) = -\frac{e}{4\pi}\frac{\varphi'\,(\tau - \tau_R)}{R}\left(\frac{c_5}{c} + \frac{c}{c_5}\right). \tag{4.20}$$

Because we obtained $f(x,\tau)$ using only the leading term $G_{Maxwell}$ in the Green's function, we expect errors on the order of the neglected term $G_{Correlation}$. In particular, we notice that

$$\left(\partial_\mu \partial^\mu + \frac{\eta_{55}}{c_5^2}\partial_\tau^2\right)G_{Maxwell} = -\delta^4\,(x)\,\delta\,(\tau) - \frac{1}{2\pi}\frac{\eta_{55}}{c_5^2}\delta(x^2)\,\delta''(\tau),$$

where the second term on the right is canceled when $G_{Correlation}$ is included in the wave equation. As a result, calculating

$$\nabla \cdot \boldsymbol{\epsilon} = \frac{c_5}{c}\frac{e}{4\pi}\left(\varphi\,(\tau - \tau_R)\,\delta^3\,(\mathbf{x}) - \frac{\varphi''\,(\tau - \tau_R)}{cR}\right),$$

where we use $\nabla \cdot (\hat{\mathbf{x}}/R^2) = 4\pi\delta^3(\mathbf{x})$, and

$$\frac{1}{c}\frac{\partial}{\partial t}\epsilon^0 = \frac{e}{4\pi}\frac{\varphi''\,(\tau - \tau_R)}{cR}\left(\frac{c_5}{c} + \frac{c}{c_5}\right)$$

leads to the Gauss law as

$$\frac{1}{c}\frac{\partial}{\partial t}\epsilon^0 + \nabla \cdot \boldsymbol{\epsilon} = \frac{c_5}{c}\frac{e}{4\pi}\varphi\,(\tau - \tau_R)\,\delta^3\,(\mathbf{x}) + \frac{c}{c_5}\frac{e}{4\pi}\frac{\varphi''\,(\tau - \tau_R)}{cR}$$

exposing an error at the order of $\delta''(\tau - \tau_R)$.

We now consider a long straight charged line oriented along the z-axis, with charge per unit length λ_e. In cylindrical coordinates

$$\mathbf{x} = (\rho, z) \qquad \rho = (x, y) = \rho\hat{\rho} \qquad \rho = \sqrt{x^2 + y^2}$$

the fields $\boldsymbol{\epsilon}$ and **e** are found by replacing $R = \sqrt{\rho^2 + z^2}$ in (4.19) and (4.20) and integrating along the z-axis to find

$$\mathbf{e} = \frac{\lambda_e}{4\pi}\int dz\left(\frac{\varphi\left(\tau - t + \frac{(\rho^2+z^2)^{1/2}}{c}\right)}{(\rho^2 + z^2)^{3/2}} - \frac{\varphi'\left(\tau - t + \frac{(\rho^2+z^2)^{1/2}}{c}\right)}{c\,(\rho^2 + z^2)}\right)(\rho\hat{\rho}, z)$$

$$\epsilon^0 = -\frac{\lambda_e}{4\pi}\frac{c_5}{c}\int dz\frac{\varphi'\left(\tau - t + \frac{(\rho^2+z^2)^{1/2}}{c}\right)}{c\,(\rho^2 + z^2)^{1/2}}.$$

To get a sense of these expressions, we may use (3.15) to approximate $\varphi(x) = \lambda\delta(x)$ which permits us to easily carry out the z-integration to obtain

$$\mathbf{e} = \frac{\lambda\lambda_e}{2\pi}\left(\frac{\theta\left(t - \rho/c - \tau\right)\rho}{c\left((t-\tau)^2 - \rho^2/c^2\right)^{3/2}} - \frac{\delta\left(t - \rho/c - \tau\right)}{\sqrt{(t-\tau)^2 - \rho^2/c^2}}\right)\hat{\rho}$$

which vanishes $\tau > \tau_R = t - \rho/c$ as required for retarded causality. Since

$$\int \frac{d\tau}{\lambda}\,\varphi(\tau) = 1 \qquad\qquad \int \frac{d\tau}{\lambda}\,\varphi'(\tau) = 0$$

the concatenated electric field is found as

$$\mathbf{E}(x) = \int \frac{d\tau}{\lambda}\,\mathbf{e}(x,\tau) = \frac{\lambda_e}{4\pi}\int dz\frac{1}{(\rho^2 + z^2)^{3/2}}\,(\rho\hat{\rho}, z) = \frac{\lambda_e}{2\pi\rho}\,(\hat{\rho}, 0)$$

in agreement with the standard expression.

To obtain the field of a charged sheet in the $x - y$ plane with charge per unit area σ, it is convenient to start from the potential from a charged event, and integrating over x and y with $R = \sqrt{x^2 + y^2 + z^2}$. Thus,

$$a^0(x,\tau) = \frac{\sigma c}{4\pi}\int dx' dy'\,\frac{\varphi\left(\tau - t + \frac{1}{c}\sqrt{(x-x')^2 + (y-y')^2 + z^2}\right)}{c\sqrt{(x-x')^2 + (y-y')^2 + z^2}}$$

and $a^5(x,\tau) = (c_5/c)a^0(x,\tau)$. Changing to radial coordinates $(x, y) \to (\rho, \theta)$ we obtain

$$a^0(x,\tau) = \frac{\sigma c}{4\pi}\int d\theta d\rho\,\frac{\varphi\left(\tau - t + \frac{1}{c}\sqrt{\rho^2 + z^2}\right)}{c\sqrt{\rho^2 + z^2}}$$

which by change of variable $\zeta = \frac{1}{c}\sqrt{\rho^2 + z^2}$ becomes

$$a^0(x,\tau) = \frac{\sigma c}{2}\int_{|z|/c}^{\infty} \varphi\left(\tau - t + \zeta\right)d\zeta.$$

We calculate the fields from

$$\mathbf{e}(x,\tau) = -\nabla a^0 = \frac{\sigma}{2}\varphi\left(\tau - t + \frac{|z|}{c}\right)\nabla|z| = \frac{\sigma}{2}\varepsilon(z)\varphi\left(\tau - t + \frac{|z|}{c}\right)\hat{\mathbf{z}},$$

where $\epsilon(x,\tau) = (c_5/c)\mathbf{e}(x,\tau)$ and

$$\epsilon^0 = \frac{\eta_{55}}{c_5}\partial_\tau a^0 + \frac{1}{c}\partial_t a^5 = \frac{1}{c}\left(\eta_{55}\frac{c}{c_5} - \frac{c_5}{c}\right)\partial_\tau a^0$$

$$= \frac{\sigma}{c}\left(\eta_{55}\frac{c}{c_5} - \frac{c_5}{c}\right)\varphi\left(\tau - t + \frac{|z|}{c}\right).$$

By concatenation, we recover

$$\mathbf{E}(x) = \int d\tau \, \mathbf{e}(x,\tau) = \int d\tau \, \frac{\sigma}{2}\varepsilon(z)\varphi\left(\tau - t + \frac{|z|}{c}\right)\hat{\mathbf{z}} = \frac{\sigma}{2}\varepsilon(z)\,\hat{\mathbf{z}}$$

in agreement with the Maxwell field from a charged sheet. We notice that, as expected, the space part of the electric fields change sign at the plane of the sheet, pointing out at each side. Consequently, an event passing through a charged sheet of equal sign will decelerate in space on its approach and then accelerate as it retreats. However, unlike the field of a point event, the temporal part ϵ^0 is an even function of spatial distance and so the event may accelerate along the time axis on both its approach to the charged sheet and its retreat. In such a case, the spatial motion will asymptotically return to its initial condition, while the event acquires a net temporal acceleration, corresponding to a shift in energy and mass.

4.4   PLANE WAVES

From the wave equation (3.22) for $j^\alpha(x,\tau) = 0$ we may write the field in terms of the Fourier transform [6]

$$f^{\alpha\beta}(x,\tau) = \frac{1}{(2\pi)^5}\int d^5k \, e^{ik\cdot x} f^{\alpha\beta}(k) = \frac{1}{(2\pi)^5}\int d^4k \, d\kappa \, e^{i(\mathbf{k}\cdot\mathbf{x}+k_0 x^0+\eta_{55}c_5\kappa\tau)} f^{\alpha\beta}(k,\kappa),$$

where

$$\kappa = k^5 = \eta^{55}k_5$$

is understood to represent the mass carried by the plane wave, much as k^0 and $\mathbf{k}$ represent energy and 3-momentum. This interpretation is supported by the wave equation which imposes the 5D constraint

$$k^\alpha k_\alpha = \mathbf{k}^2 - (k^0)^2 + \eta_{55}\kappa^2 = 0 \quad \Longrightarrow \quad \eta_{55}\kappa^2 = (k^0)^2 - \mathbf{k}^2 \qquad (4.21)$$

expressing κ in terms of the difference between energy and momentum. Under concatenation, the field becomes

$$F^{\alpha\beta}(x) = \int \frac{d\tau}{\lambda} f^{\alpha\beta}(x,\tau) = \int \frac{d^4k}{(2\pi)^4} e^{ik_\mu x^\mu}\frac{1}{\lambda c_5}f^{\alpha\beta}(k,0) = \int \frac{d^4k}{(2\pi)^4} e^{ik_\mu x^\mu} F^{\alpha\beta}(k)$$

and recovers the 4D mass-shell constraint $k^\mu k_\mu = 0$ for the Maxwell field. In the transform domain, the sourceless pre-Maxwell equations take the form

$$\mathbf{k}\cdot\mathbf{e} - \eta_{55}\kappa\epsilon^0 = 0 \qquad\qquad \mathbf{k}\cdot\mathbf{b} = 0 \qquad\qquad \mathbf{k}\cdot\boldsymbol{\epsilon} - k^0\epsilon^0 = 0$$

$$\mathbf{k}\times\mathbf{e} - k^0\mathbf{b} = 0 \qquad\qquad\qquad\qquad \mathbf{k}\times\mathbf{b} + k^0\mathbf{e} - \eta_{55}\kappa\boldsymbol{\epsilon} = 0$$

$$\mathbf{k}\times\boldsymbol{\epsilon} - \kappa\mathbf{b} = 0 \qquad\qquad\qquad\qquad -\kappa\mathbf{e} + k^0\boldsymbol{\epsilon} - \mathbf{k}\epsilon^0 = 0$$

which can be solved by taking $\epsilon_\parallel$ and $e_\perp$ as independent 3-vector polarizations, and writing

$$\mathbf{e}_\parallel = \eta_{55}\frac{\kappa}{k^0}\boldsymbol{\epsilon}_\parallel \qquad \boldsymbol{\epsilon}_\perp = \frac{\kappa}{k^0}\mathbf{e}_\perp \qquad \epsilon^0 = \frac{1}{k^0}\mathbf{k}\cdot\boldsymbol{\epsilon}_\parallel \qquad \mathbf{b} = \frac{1}{k^0}\mathbf{k}\times\mathbf{e}_\perp$$

for the remaining fields. Unlike Maxwell plane waves, for which $\mathbf{E}$, $\mathbf{B}$, and $\mathbf{k}$ are mutually orthogonal, the pre-Maxwell electric fields $\mathbf{e}$ and $\boldsymbol{\epsilon}$ have both transverse and longitudinal components. When $\kappa \to 0$, we find that $\mathbf{e}$, $\mathbf{b}$, and $\mathbf{k}$ become mutually orthogonal and $\boldsymbol{\epsilon}$ becomes a decoupled longitudinal polarization parallel to $\mathbf{k}$.

We use (3.11) to write the convolved field as

$$f_\Phi^{\alpha\beta}(x,\tau) = \int \frac{ds}{\lambda}\,\Phi(\tau-s)f^{\alpha\beta}(x,s) = \frac{1}{(2\pi)^5}\int d^4k\,d\kappa\,e^{i(\mathbf{k}\cdot\mathbf{x}-k_0x^0+\eta_{55}c_5\kappa\tau)}f_\Phi^{\alpha\beta}(k,\kappa),$$

where

$$f_\Phi^{\alpha\beta}(k,\kappa) = \frac{1+(\xi\lambda c_5\kappa)^2}{\lambda}f^{\alpha\beta}(k,\kappa)$$

introduces a multiplicative factor that will appear once in each field bilinear of $T_\Phi^{\alpha\beta}$. In terms of the 3-vector fields, the mass-energy-momentum tensor components are

$$T_\Phi^{00} = \frac{1}{2c}\left[\mathbf{e}\cdot\mathbf{e}_\Phi + \mathbf{b}\cdot\mathbf{b}_\Phi + \eta_{55}\left(\boldsymbol{\epsilon}\cdot\boldsymbol{\epsilon}_\Phi + \epsilon^0\epsilon_\Phi^0\right)\right]$$

$$T_\Phi^{0i} = \frac{1}{c}\left(\mathbf{e}\times\mathbf{b}_\Phi + \eta_{55}\epsilon^0\boldsymbol{\epsilon}_\Phi\right)^i$$

$$T_\Phi^{50} = \frac{1}{c}\mathbf{e}\cdot\boldsymbol{\epsilon}_\Phi$$

$$T_\Phi^{5i} = \frac{1}{c}\left(\boldsymbol{\epsilon}\times\mathbf{b}_\Phi + \epsilon^0\mathbf{e}_\Phi\right)^i$$

$$T_\Phi^{55} = \frac{1}{2c}\left[\boldsymbol{\epsilon}\cdot\boldsymbol{\epsilon}_\Phi - \epsilon^0\epsilon_\Phi^0 + \eta^{55}\left(\mathbf{e}\cdot\mathbf{e}_\Phi - \mathbf{b}\cdot\mathbf{b}_\Phi\right)\right].$$

For the plane wave, the energy density is

$$T_\Phi^{00} = \frac{1}{c}\left(\mathbf{e}_\perp^2 + \eta_{55}\boldsymbol{\epsilon}_\parallel^2\right)\frac{1+(\xi\lambda\kappa)^2}{\lambda}$$

which, since $\mathbf{e}_\perp^2 = \frac{1}{2}\left(\mathbf{e}_\perp^2 + \mathbf{b}^2\right)$, is equivalent in form to the energy density in Maxwell theory

$$\theta^{00} = \frac{1}{2c}\left(\mathbf{E}^2 + \mathbf{B}^2\right)$$

with the addition of the independent polarization $\boldsymbol{\epsilon}_\parallel$. The mass density is found to be

$$T_\Phi^{55} = \frac{\kappa^2}{ck_0^2}\left(\mathbf{e}_\perp^2 + \eta_{55}\boldsymbol{\epsilon}_\parallel^2\right)\frac{1+(\xi\lambda\kappa)^2}{\lambda} = \frac{\kappa^2}{k_0^2}T_\Phi^{00}$$

expressing energy density scaled by the squared mass-to-energy ratio for the field. The energy flux—the standard Poynting 3-vector—is

$$T_\Phi^{0i} \longrightarrow \mathbf{T}_\Phi^0 = \frac{\mathbf{k}}{k^0} T_\Phi^{00}$$

expressing the energy density T_Φ^{00} flowing uniformly in the direction of the momentum normalized to energy. Comparing the proportionality factor to that for a free particle

$$\frac{\mathbf{k}}{k^0} \longrightarrow \frac{\mathbf{p}}{E/c} = \frac{1}{c}\frac{M\,d\mathbf{x}/d\tau}{M\,dt/d\tau} = \frac{\mathbf{v}}{c}$$

which will not generally be a unit vector unless $\kappa = 0$, as it must be for Maxwell plane waves. The mass flux vector—a second Poynting 3-vector—can be written

$$T_\Phi^{5i} \longrightarrow \mathbf{T}_\Phi^5 = \frac{\mathbf{k}}{\kappa} T_\Phi^{55}$$

expressing the mass density T_Φ^{55} flowing uniformly in the direction of the momentum normalized to mass. Finally,

$$T_\Phi^{50} = \frac{k^0}{\kappa} T_\Phi^{55} = \frac{\kappa}{k^0} T_\Phi^{00} \qquad .$$

so that $T_\Phi^{5\mu}$ can be written as

$$T_\Phi^{5\mu} = \frac{k^\mu}{\kappa} T_\Phi^{55} = \frac{\kappa k^\mu}{k_0^2} T_\Phi^{00}$$

expressing the mass density T_Φ^{55} flowing in the direction of the 4-momentum. In this sense, T_Φ^{50} represents the flow of mass into the time direction. We notice that when $\kappa \longrightarrow 0$, as is the case for Maxwell plane waves, $\mathbf{k}/k^0$ becomes a unit vector and $T_\Phi^{5\alpha} = 0$, so that mass density and flow vanish. The interpretation of plane waves carrying energy and momentum (energy flux) uniformly to infinity is thus seen to generalize to mass flow, where mass is best understood through (4.21) as the non-identity of energy and momentum.

Suppose that a plane wave of this type impinges on a test particle in its rest frame, described by $x^\alpha(\tau) = (c\tau, 0, c_5\tau)$. Since $\dot{\mathbf{x}} = 0$, the wave will interact with the event through the Lorentz force (3.6) and (3.7) as

$$M\ddot{x}^\mu(\tau) = \frac{e}{c}\left[f^\mu_{\ 0}(x,\tau)\dot{x}^0(\tau) + c_5 f^\mu_{\ 5}(x,\tau) \right] \qquad \frac{d}{d\tau}\left(-\tfrac{1}{2}M\dot{x}^2\right) = -\frac{ec_5}{c}\eta_{55}\epsilon^0\dot{x}^0$$

which for $\epsilon_\parallel \neq 0 \Rightarrow \mathbf{k}\cdot\epsilon_\parallel \neq 0$ becomes

$$\ddot{t} = -\eta_{55}e\frac{c_5}{Mc^2}\frac{1}{k^0}\mathbf{k}\cdot\epsilon_\parallel \qquad \ddot{\mathbf{x}} = \frac{e}{M}\left[\mathbf{e}_\perp\left(1 + \frac{c_5}{c}\frac{\kappa}{k^0}\right) + \epsilon_\parallel\left(\frac{c_5}{c} + \eta_{55}\frac{\kappa}{k^0}\right) \right]$$

$$\frac{d}{d\tau}\left(-\tfrac{1}{2}M\dot{x}^2\right) = -\eta_{55}ec_5\frac{1}{k^0}\mathbf{k}\cdot\boldsymbol{\epsilon}_{\parallel}$$

showing that the incident plane wave will initially accelerate the test event in such a way as to transfer mass. If the plane wave is a far field approximation to the radiation field of an accelerating charge, then the resulting picture describes the transfer of mass by the radiation field between charged events.

4.5   RADIATION FROM A LINE ANTENNA

The radiation from a dipole antenna is treated generally in Maxwell theory [7] by approximating the oscillating current as the separable current density

$$\mathbf{J}(\mathbf{x},t) = \mathbf{J}(\mathbf{x})\,e^{i\omega t} \qquad\qquad \nabla\cdot\mathbf{J}(\mathbf{x}) + i\omega\rho = 0,$$

where the second equation expresses represents current conservation, and of course we take the real parts of all physical quantities. This approximation may be justified by posing a collection of oscillating charges with position 4-vectors

$$X_n(\tau) = \left(ct_n(\tau),\ \mathbf{a}_n e^{i\omega\tau}\right)$$

which for nonrelativistic motion includes $t_n(\tau) = t = \tau$ for each particle. The Maxwell current for this collection is

$$\mathbf{J}(\mathbf{x},t) = \sum_n \int d\tau\, c\dot{\mathbf{X}}_n(\tau)\,\delta^4\left(x - X_n(\tau)\right)$$

$$= \sum_n \int d\tau\, i\omega\, c\mathbf{a}_n e^{i\omega\tau}\delta\left(ct - c\tau\right)\delta^3\left(\mathbf{x} - \mathbf{a}_n e^{i\omega\tau}\right)$$

$$= \left[\sum_n i\omega\, \mathbf{a}_n\delta^3\left(\mathbf{x} - \mathbf{a}_n e^{i\omega t}\right)\right]e^{i\omega t}$$

so that replacing the term in square brackets with its time average over one cycle of oscillation $T = 2\pi/\omega$ we obtain

$$\mathbf{J}(\mathbf{x},t) \simeq \left[\frac{1}{T}\int_0^T dt\,\sum_n i\omega\, \mathbf{a}_n\delta^3\left(\mathbf{x} - \mathbf{a}_n e^{i\omega t}\right)\right]e^{i\omega t} = \mathbf{J}(\mathbf{x})\,e^{i\omega t}.$$

Thus, $\mathbf{J}(\mathbf{x})$ approximates the time-dependent current density by a time averaged static configuration in space, rendering the antenna problem tractable.

To treat the dipole antenna in SHP electrodynamics [8] we cannot make use of this approximation because the microscopic current

$$\mathbf{j}(\mathbf{x},t,\tau) = c\sum_n \dot{\mathbf{X}}_n(\tau)\,\delta^4\left(x - X_n(\tau)\right) = \sum_n i\omega\, \mathbf{a}_n e^{i\omega\tau}\delta\left(t - \tau\right)\delta^3\left(\mathbf{x} - \mathbf{a}_n e^{i\omega\tau}\right)$$

is not integrated over τ, and so time averaging cannot be performed in any meaningful way. Instead, in analogy to this approximation, we pose a current of the form

$$j^0(x, \tau) = c\left[\rho_0(x) + \rho(x)e^{i\omega t}\right]\phi(\tau - t)$$

$$\mathbf{j}(x, \tau) = \mathbf{J}(x)e^{i\omega t}\phi(\tau - t)$$

$$j^5(x, \tau) = \frac{c_5}{c}j^0(x, \tau) = c_5\left[\rho_0(x) + \rho(x)e^{i\omega t}\right]\phi(\tau - t),$$

where $\rho_0(x)$ is a background event density. The function $\phi(\tau - t)$ expresses a correlation between t and τ, inserted by hand in place of a time averaging procedure. In this sense, the replacement

$$\mathbf{j}(x, t, \tau) \quad \longrightarrow \quad \mathbf{J}(x)e^{i\omega t}\phi(\tau - t)$$

may be less precise than the comparable approximation in Maxwell theory, and we must be attentive to artifacts introduced by the model. In analogy to (3.15), we choose

$$\phi(\tau - t) = \frac{1}{2\sigma}e^{-|\tau - t|/\sigma} = \int \frac{d\omega}{2\pi}\Phi(\omega)e^{i\omega(\tau - t)} \qquad \Phi(\omega) = \frac{1}{1 + (\sigma\omega)^2}$$

which imposes a correlation $\tau - t \simeq \sigma$ through

$$\phi(\tau - t) \longrightarrow \begin{cases} \text{strong correlation:} & \sigma \to 0 \implies \phi(\tau - t) \to \delta(\tau - t) \implies t = \tau \\ \text{weak correlation:} & \sigma \to \text{large} \implies t - \tau \text{ evenly distributed.} \end{cases}$$

Notice that in the strong correlation limit, the potential found from the Green's function

$$\mathbf{a}(x, \tau) = \frac{e}{2\pi}\int d^3x' d(ct')\,\delta\left((x - x')^2 - c^2(t - t')^2\right)\mathbf{J}(x')e^{i\omega t'}\delta(\tau - t')$$

$$= \frac{e}{4\pi c}e^{i\omega\tau}\int d^3x'\,\frac{1}{|x - x'|}\delta\left(\tau - t + \frac{|x - x'|}{c}\right)\mathbf{J}(x')$$

describes a Coulomb-like potential oscillating in τ simultaneously across spacetime, rather than a wave propagating with phase $(kr - \omega t)$. This suppression of the expected wavelike behavior can be characterized by the dimensionless parameter

$$\frac{1}{\omega\sigma} = \frac{T}{2\pi\sigma} = \frac{\text{antenna period}}{\text{correlation time}}$$

which we take to be small but greater than zero. The total number of events in this system at time τ is found from the spacetime integral

$$N(\tau) = \frac{1}{c_5} \int d^4x \, j^5(x, \tau)$$

$$= \int d^3x \, \rho_0(\mathbf{x}) \int dt \, \phi(\tau - t) + \int d^3x \, \rho(\mathbf{x}) \int dt \, e^{i\omega t} \phi(\tau - t)$$

$$= N_0 + \frac{N e^{i\omega\tau}}{1 + (\sigma\omega)^2},$$

where

$$N_0 = \int d^3x \, \rho_0(\mathbf{x}) \qquad\qquad N = \int d^3x \, \rho(\mathbf{x})$$

given as a background event number with an oscillating perturbation. We must have

$$N_0 > \frac{N}{1 + (\sigma\omega)^2}$$

to unsure that the event number remains positive. Similarly, the total charge is given by

$$Q(\tau) = Q_0 + \frac{Q e^{i\omega\tau}}{1 + (\sigma\omega)^2},$$

where $Q_0 = e N_0$ and $Q = e N$, so that the total charge does not change sign, which would suggest pair creation and annihilation processes. Since the background density $\rho_0(\mathbf{x})$ is independent of t and τ, conservation of the 5D current becomes

$$0 = \frac{1}{c} \frac{\partial}{\partial t} j^0 + \nabla \cdot \mathbf{j} + \frac{1}{c_5} \frac{\partial}{\partial \tau} j^5 = \rho(\mathbf{x}) \frac{\partial}{\partial t} \left[\left(e^{i\omega t}\right) \phi(\tau - t) \right] + \nabla \cdot \mathbf{J}(\mathbf{x}) e^{i\omega t} \phi(\tau - t)$$

$$+ \rho(\mathbf{x}) e^{i\omega t} \frac{\partial}{\partial \tau} \phi(\tau - t)$$

$$= \left[i\omega \rho(\mathbf{x}) + \nabla \cdot \mathbf{J}(\mathbf{x}) \right] e^{i\omega t} \phi(\tau - t)$$

so that

$$i\omega \rho + \nabla \cdot \mathbf{J} = 0 \longrightarrow e \int d^3x \, \mathbf{J}(\mathbf{x}) = -e \int d^3x \, \mathbf{x} \, \nabla \cdot \mathbf{J} = e \int d^3x \, \mathbf{x} \, (i\omega\rho)$$

and we identify

$$\mathbf{p} = \int d^3x \, \mathbf{x} \, e \rho(\mathbf{x}) \qquad\qquad i\omega\mathbf{p} = I d \hat{\mathbf{d}}$$

as the dipole moment $\mathbf{p}$ of the charge distribution $\rho(\mathbf{x})$, so that $i\omega\mathbf{p}$ can be written as a constant current I along a dipole of length d in the direction $\hat{\mathbf{d}}$. The total current density is

$$\mathbf{J}(\tau) = \frac{e}{c} \int d^4x \, \mathbf{J}(\mathbf{x}) e^{i\omega t} \phi(\tau - t) = \frac{i\omega\mathbf{p} e^{i\omega\tau}}{1 + (\sigma\omega)^2}$$

representing an oscillating dipole.

The induced potential found from the Green's function $G_{Maxwell}$ is

$$a^\alpha(x,\tau) = \frac{e}{c}\int d^3x' \frac{1}{4\pi|x-x'|} j^\alpha\left(c\left(t - \frac{|x-x'|}{c}\right), x', \tau\right)$$

so that writing $x = r\hat{r}$, we make the far field approximation

$$R = |x-x'| = \left(r^2 + (x')^2 - 2r\hat{r}\cdot x'\right)^{1/2} \simeq r - \hat{r}\cdot x'$$

and the dipole approximation

$$|k\hat{r}\cdot x'| < kd = \frac{2\pi d}{\lambda} \ll 1 \Rightarrow e^{ik\hat{r}\cdot x'} \simeq 1 \qquad \frac{r - \hat{r}\cdot x'}{c} \simeq \frac{r}{c}\left(1 - \hat{r}\cdot\hat{x}'\frac{d}{r}\right) \simeq \frac{r}{c}$$

to obtain

$$a^0(x,\tau) \simeq \frac{Q_0}{4\pi r} + \frac{Q}{4\pi r}e^{-i(kr-\omega t)}\phi\left(\tau - t + \frac{r}{c}\right)$$

$$a(x,\tau) \simeq p\frac{ik}{4\pi r}e^{-i(kr-\omega t)}\phi\left(\tau - t + \frac{r}{c}\right)$$

$$a^5(x,\tau) = \frac{c_5}{c}a^0(x,\tau).$$

We define the spherical wave factor

$$\chi(x,\tau) = \frac{e^{-i(kr-\omega t)}}{4\pi r}\phi\left(\tau - t + \frac{r}{c}\right)$$

and split the field strengths into spacetime and polarization factors, as

$$b = \nabla\times a = \hat{b}\chi \qquad\qquad \hat{b} = -ikId\varepsilon_1\,\hat{r}\times\hat{d}$$

$$e = -\frac{1}{c}\frac{\partial}{\partial t}a - \nabla a^0 = \frac{Q_0}{4\pi r^2}\hat{r} + \hat{e}\chi \qquad \hat{e} = ik\varepsilon_1\left(Q\hat{r} - Id\hat{d}\right)$$

$$\epsilon = \eta_{55}\frac{1}{c_5}\frac{\partial}{\partial\tau}a - \frac{c_5}{c}\nabla a^0 = \frac{c_5}{c}\frac{Q_0}{4\pi r^2}\hat{r} + \hat{\epsilon}\chi \qquad \hat{\epsilon} = ik\left[\frac{c_5}{c}\varepsilon_1 Q\hat{r} + i\varepsilon_2 Id\hat{d}\right]$$

$$\epsilon^0 = \eta_{55}\frac{1}{c_5}\frac{\partial}{\partial\tau}a^0 - \frac{1}{c}\frac{\partial}{\partial t}a^5 = \hat{\epsilon}^0\chi \qquad \hat{\epsilon}^0 = ik\left[\frac{c_5}{c}\varepsilon_1 + i\varepsilon_2\right]Q,$$

where we used $1/kr \ll 1$ and define

$$\varepsilon_1 = 1 + \frac{\varepsilon(\tau - t + R/c)}{i\omega\sigma} \qquad\qquad \varepsilon_2 = -\eta_{55}\frac{c}{c_5}\frac{\varepsilon(\tau - t + R/c)}{\omega\sigma}.$$

We drop the static Coulomb terms produced by Q_0, as these do not contribute to radiation. Since $\omega\sigma \sim$ small tends to suppress wavelike behavior, but $c_5/c \ll 1$ [9], we approximate $\varepsilon_1 \simeq 1$ but leave ε_2 unchanged. Taking the orientation of the antenna to be $\hat{\mathbf{d}} = \hat{\mathbf{z}}$ the polarizations then simplify to

$$\hat{e} \simeq ik\,(Q\hat{\mathbf{r}} - Id\hat{\mathbf{z}}) \qquad \hat{b} \simeq -ikId\,\hat{\mathbf{r}} \times \hat{\mathbf{z}}$$

$$\hat{\epsilon}^0 \simeq ik\left[\frac{c_5}{c} + i\varepsilon_2\right]Q \qquad \hat{\epsilon} \simeq ik\left[\frac{c_5}{c}Q\hat{\mathbf{r}} + i\varepsilon_2 Id\hat{\mathbf{z}}\right]$$

and we notice that terms containing $1/\omega\sigma$ appear only in the components of $\hat{\epsilon}^\mu$. Such terms are artifacts of modeling the time correlation by $\phi(\tau - t)$, and can be understood as the contribution to the fields required to impose this correlation across spacetime. As was seen for plane waves, these fields will accelerate a test event initially at rest through the Lorentz force in such a way as to transfer mass to the event.

The mass-energy-momentum tensor will contain bilinear field combinations of the type

$$T^{\alpha\beta} = \frac{1}{c}\left(f_\Phi^{\alpha\gamma} f^\beta{}_\gamma - \frac{1}{4}f_\Phi^{\delta\varepsilon} f_{\delta\varepsilon} g^{\alpha\beta}\right) \longrightarrow \mathrm{Re}\left[\left(\mathbf{A}^\alpha + i\,\mathbf{B}^\alpha\right)\chi\right] \cdot \mathrm{Re}\left[\left(\mathbf{C}^\beta + i\,\mathbf{D}^\beta\right)\chi\right]$$

and it is convenient to separate the resulting products as $T^{\alpha\beta} = T_0^{\alpha\beta} + T_\sigma^{\alpha\beta}$, where $T_\sigma^{\alpha\beta}$ includes all terms containing $1/\omega\sigma$. We designate

$$S(x,\tau) = k^2\left(\frac{\phi\left(\tau - t + \frac{r}{c}\right)}{4\pi r}\right)^2 \sin^2(kr - \omega t)$$

$$C(x,\tau) = k^2\left(\frac{\phi\left(\tau - t + \frac{r}{c}\right)}{4\pi r}\right)^2 \cos^2(kr - \omega t)$$

$$X(x,\tau) = k^2\left(\frac{\phi\left(\tau - t + \frac{r}{c}\right)}{4\pi r}\right)^2 2\sin(kr - \omega t)\cos(kr - \omega t)$$

and note that these functions drop off as $1/r^2$ and so will produce nonzero surface integrals at large r, as is characteristic of radiation fields. Using these functions, the components of $T_0^{\alpha\beta}$ are

$$T_0^{00} = \frac{1}{2}\left[(Q + Id\,\cos\theta)^2 + 2(Id)^2\left(1 - (\cos\theta)^2\right) + 2\eta_{55}\left(\frac{c_5}{c}\right)^2 Q^2\right]S(x,\tau)$$

$$\mathbf{T}_0^0 = \left[Id\,(Q - Id\,\cos\theta)\hat{\mathbf{z}} + \left(Id\,(Id - Q\,\cos\theta) + \eta_{55}\left(Q\frac{c_5}{c}\right)^2\right)\hat{\mathbf{r}}\right]S(x,\tau)$$

$$T_0^{50} = Q\frac{c_5}{c}(Q - Id\,\cos\theta)\,S(x,\tau)$$

$$\mathbf{T}_0^5 = Q\frac{c_5}{c}(Q - Id\,\cos\theta)\,S(x,\tau)\hat{\mathbf{r}} = T_0^{50}\,\hat{\mathbf{r}}$$

$$T_0^{55} = \frac{1}{2}\eta^{55}(Q - Id\,\cos\theta)^2\,S(x,\tau)$$

all of which have spacetime dependence $S(x, \tau)$. The components of $T_\sigma^{\alpha\beta}$ are

$$T_\sigma^{00} = \frac{1}{2}\left[\varepsilon_2^2\left[(Id)^2 + Q^2\right]C(x,\tau) - \eta_{55}\frac{c_5}{c}\varepsilon_2 Q\left[Q + Id\,\cos\theta\right]X(x,\tau)\right]$$

$$\mathbf{T}_\sigma^0 = -\eta_{55}\frac{c_5}{c}\varepsilon_2 Q\left[Id\left(X(x,\tau) + \eta_{55}\frac{c}{c_5}C(x,\tau)\right)\hat{z} + Q\,X(x,\tau)\hat{r}\right]$$

$$T_\sigma^{50} = -\varepsilon_2 Id\,(Id - Q\,\cos\theta)\,X(x,\tau)$$

$$\mathbf{T}_\sigma^5 = -\varepsilon_2\left[Id\left[Q - Id\,\cos\theta\right]\hat{z} + \left((Id)^2 - Q^2\right)\hat{r}\right]X(x,\tau)$$

$$T_\sigma^{55} = \frac{1}{2}\left[\varepsilon_2^2\left((Id)^2 - Q^2\right)C(x,\tau) + \frac{c_5}{c}\varepsilon_2 Q\,(Q - Id\,\cos\theta)\,X(x,\tau)\right]$$

whose spacetime dependence is determined by $C(x, \tau)$ and $X(x, \tau)$ and is thus out of phase with the $T_0^{\alpha\beta}$.

As expected from the transfer of mass made possible by the fields ϵ^μ, we find a nonzero mass density T^{55} and mass flux $\mathbf{T}^0$ and $\mathbf{T}^5$ into time and space. Moreover, integrating over a sphere of radius r, the net mass flux into space will be of the form

$$P = \int d\Omega\, r^2\,\hat{r}\cdot\mathbf{T}_0^5$$

$$= \int d\Omega\, r^2\,\hat{r}\cdot\left[Q\frac{c_5}{c}(Q - Id\,(\cos\theta))\,S(x,\tau)\hat{r}\right]$$

$$= Q\frac{c_5}{c}r^2\,k^2\left(\frac{\phi\left(\tau - t + \frac{r}{c}\right)}{4\pi r}\right)^2\sin^2(kr - \omega t)\int d\Omega\,[Q - Id\,\cos\theta]$$

$$= \frac{k^2 c_5}{4\pi c}Q^2\left(\phi\left(\tau - t + \frac{r}{c}\right)\right)^2\sin^2(kr - \omega t)$$

and thus nonzero wherever $\tau \simeq t - r/c$. Just as the energy radiated by a Maxwell dipole antenna must be provided by the amplifier that drives the oscillating current density, the mass radiated by an SHP antenna is continuously provided by an amplifier that creates events and drives them into the antenna.

For a center-fed antenna of length d oriented along the z-axis, the charge density may be described by

$$\rho(\mathbf{x}) = \begin{cases} \delta(x)\,\delta(y)\,\rho_z(z) & , \quad -\frac{d}{2} \le z \le \frac{d}{2} \\ 0 & , \quad \text{otherwise}, \end{cases}$$

where

$$\rho_z(z) = \frac{1}{2}\left[\rho_z(z) + \rho_z(-z)\right] + \frac{1}{2}\left[\rho_z(z) - \rho_z(-z)\right] = \rho_+(z) + \rho_-(z)$$

divides the charge density into even and odd parts. The total oscillating charge is

$$Q = \int d^3x \, e\rho(\mathbf{x}) = 2e \int_0^{\frac{d}{2}} dz \, \rho_+(z)$$

and the dipole moment is

$$Id \, \hat{\mathbf{d}} = i\omega e \int d^3x \, \mathbf{x}\rho(\mathbf{x}) = 2ie\omega\hat{\mathbf{z}} \int_0^{\frac{d}{2}} dz \, z \, \rho_-(z)$$

showing that ρ_+ describes a net charge Q driven symmetrically into the left and right segments of the antenna, while ρ_- describes a dipole moment produced by shifting charge from one antenna segment into the other segment. Since $Q = eN$, we see that the amplifier driving net charge into the antenna must be driving new events into the antenna as well, accounting for the radiated mass. Taking $Q = 0$ so that the amplifier shifts charged events between antenna segments without injecting new events, the fields reduce to

$$\hat{e} = -ikId\hat{\mathbf{z}} \qquad \hat{b} = -ikId \, \hat{\mathbf{r}} \times \hat{\mathbf{z}}$$

$$\hat{\epsilon}^0 = 0 \qquad \hat{\boldsymbol{\epsilon}} = k\varepsilon_2 Id\hat{\mathbf{z}}$$

so that the effect of the waves on a test event at rest reduces to

$$\frac{d}{d\tau}\left(-\tfrac{1}{2}M\dot{x}^2\right) = -e\eta_{55}c_5\epsilon^0 = 0$$

and there is no transfer of mass. Similarly, the components of become $T_0^{\alpha\beta}$

$$T_0^{00} = (Id)^2 \left(1 - \frac{1}{2}\cos^2\theta\right) S(x,\tau)$$

$$\mathbf{T}_0^0 = (Id)^2 \left(-\cos\theta\,\hat{\mathbf{z}} + \hat{\mathbf{r}}\right) S(x,\tau)$$

$$T_0^{50} = 0$$

$$\mathbf{T}_0^5 = 0$$

$$T_0^{55} = \frac{1}{2}\eta^{55}(Id)^2\cos^2\theta\, S(x,\tau)$$

describing no transfer of mass into space or time.

The components of $T_\sigma^{\alpha\beta}$ also simplify to

$$T_\sigma^{00} = \frac{1}{2}\left(\frac{c}{\omega c_5}\frac{\phi'}{\phi}\right)^2 (Id)^2\, C\,(x,\tau)$$

$$\mathbf{T}_\sigma^0 = 0$$

$$T_\sigma^{50} = \frac{c}{\omega c_5}\eta_{55}\frac{\phi'}{\phi}\,(Id)^2\, X\,(x,\tau)$$

$$\mathbf{T}_\sigma^5 = T^{50}\,(\hat{\mathbf{z}} - \hat{\mathbf{r}})$$

$$T_\sigma^{55} = \frac{1}{2}\left(\frac{c}{\omega c_5}\frac{\phi'}{\phi}\right)^2 (Id)^2\, C\,(x,\tau),$$

where we replaced $-\varepsilon_R/\sigma$ with ϕ'/ϕ. These expressions involve no transfer of energy but do describe nonzero transfer of mass into space and time directions. Once again, we understand this transfer as an artifact of the time correlation model that enters through the derivative of $\phi\,(\tau - t)$, rather than an inherent feature of radiation from an oscillating charge. In particular, all of the nonzero terms in the expression for mass conservation contain ϕ', so that these terms are separately conserved among themselves. To see this we expand (3.25) as

$$\partial_\alpha T^{\alpha 5} = -\frac{e}{c}f^{5\alpha}j_\alpha \quad\longrightarrow\quad \frac{1}{c}\frac{\partial}{\partial t}T^{50} + \nabla\cdot\mathbf{T}^5 + \frac{1}{c_5}\frac{\partial}{\partial\tau}T^{55} = -\frac{e}{c}\epsilon^\mu j_\mu$$

which becomes

$$\frac{1}{c}\frac{\partial}{\partial t}T_\sigma^{50} + \nabla\cdot\mathbf{T}_\sigma^5 + \frac{1}{c_5}\frac{\partial}{\partial\tau}\left(T_0^{55} + T_\sigma^{55}\right) = -\frac{e}{c}\epsilon_\sigma^\mu j_\mu \qquad (4.22)$$

because $T_0^{50} = \mathbf{T}_0^5 = 0$. We also write the field as ϵ_σ^μ because it contains the factor ε_2. Finally, we note that because T_0^{55} depends on τ only through the factor of ϕ^2 in $S\,(x,\tau)$, the derivative $\partial_\tau T_0^{55}$ must similarly contain ϕ'. Thus, each term in (4.22) enters through the derivative of the time correlation model, and these terms are conserved among themselves with no corresponding energy transfer.

Integrating the energy Poynting vector $\mathbf{T}^0$ over the surface of a sphere of radius r we must evaluate

$$\hat{\mathbf{r}}\cdot\mathbf{T}^0 = (Id)^2\,\hat{\mathbf{r}}\cdot\left(-\cos\theta\,\hat{\mathbf{z}} + \hat{\mathbf{r}}\right)S\,(x,\tau)$$
$$= (Id)^2\left(-\cos^2\theta + 1\right)S\,(x,\tau)$$
$$= (Id)^2\sin^2\theta\,S\,(x,\tau)$$

to find the instantaneous radiated power

$$P = \int d\Omega \; r^2 \; (Id)^2 \, S\,(x,\tau)\sin^2\theta$$

$$= (Id)^2 \, r^2 \, k^2 \left(\frac{\phi\left(\tau - t + \frac{r}{c}\right)}{4\pi r}\right)^2 \sin^2\left(kr - \omega t\right)\int_0^{2\pi} d\phi \int_0^{\pi} d\theta \; \sin^3\theta$$

$$= (Id)^2 \, k^2 \left(\frac{\phi\left(\tau - t + \frac{r}{c}\right)}{4\pi}\right)^2 \sin^2\left(kr - \omega t\right)\frac{8\pi}{3}$$

$$= \frac{k^2 (Id)^2}{6\pi}\left(\phi\left(\tau - t + \frac{r}{c}\right)\sin\left(kr - \omega t\right)\right)^2 .$$

Since we have assumed that $1/\omega\sigma$ is small, we may take $\phi\left(\tau - t + \frac{r}{c}\right)$ as effectively constant over one cycle of the wave, so that the average radiated power over one cycle is

$$\bar{P} \simeq \frac{k^2 (Id)^2}{6\pi}\left(\phi\left(\tau - t + \frac{r}{c}\right)\right)^2 \frac{1}{T}\int_0^T dt \; (\sin\left(kr - \omega t\right))^2 = \frac{k^2 (Id)^2}{12\pi}\left(\phi\left(\tau - t + \frac{r}{c}\right)\right)^2$$

which agrees with the standard result up to the factor of ϕ^2. The neutral antenna radiates energy in agreement with the Maxwell result and radiates no mass (leaving aside the derivatives of the arbitrarily chosen function ϕ).

4.6   CLASSICAL PAIR PRODUCTION

A standard technique for pair creation in the laboratory is the two-step process by which Anderson [10] first observed positrons in 1932: high energy electrons are first scattered by heavy nuclei to produce bremsstrahlung radiation, and electron/positron pairs are then created from the radiation field. The Bethe-Heitler mechanism [11] describes this technique as the quantum process,

$$e^- + Z \longrightarrow e^- + Z + \gamma$$
$$Z + \gamma \longrightarrow Z + e^- + e^+$$

involving a quantized radiation field and the external Coulomb field of the nuclei. We now calculate the classical trajectories that produce this two-step process, as shown in Figure 4.1.

Because the electromagnetic interaction is instantaneous in τ, we may take both stages of the Bethe-Heitler process as occurring at τ_2: (1) the scattering of particle-2 by a nucleus at t_1 ($E_{in} > 0 \longrightarrow E_{out} > 0$) and (2) the absorption of the resulting bremsstrahlung radiation by particle-1 at t_2 ($E_{in} < 0 \longrightarrow E_{out} > 0$). In the Stueckelberg picture, the $E < 0$ (antiparticle) trajectory of particle-1 must have been produced at the earlier chronological time $\tau_1 < \tau_2$. To examine the conditions that might produce this initial negative energy trajectory, we describe particle-1 scattering in the Coulomb field of another nucleus at $t = t_3$ and emerging with negative energy moving backward in t.

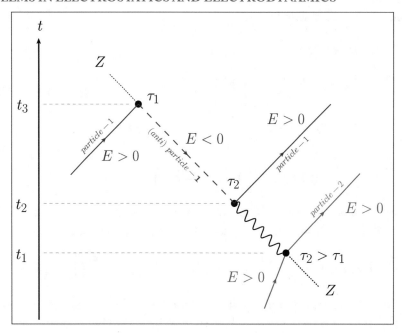

Figure 4.1: Bethe–Heitler mechanism in classical electrodynamics.

In the laboratory, where events are recorded in the order determined by clock t, the process appears as particle-2 scattering at $t = t_1$ and emitting bremsstrahlung, followed by the appearance at $t = t_2$ of a particle/antiparticle pair. Then at $t = t_3$, the antiparticle encounters another particle causing their mutual annihilation.

Our analysis is carried out in three parts. We first consider the Coulomb scattering of a slow incoming particle by an oppositely charged nucleus. To produce the pair annihilation observed at τ_1, the outgoing particle must have $E < 0$, while at τ_2 the interaction must lead to $E > 0$. We identify the condition that allows the energy of the outgoing particle to change sign. In the second part, we compute the radiation field produced by the acceleration of the scattered particle at τ_2 using the Liénard–Wiechert potential for an arbitrary trajectory. In the third part we again use the Lorentz force to treat the acceleration of the $E < 0$ particle absorbing the radiation at τ_2, and find the condition for its return to an $E > 0$ trajectory.

With the function $\varphi(\tau - \tau_1)$ in the field strengths, the Lorentz force is a set of coupled nonlinear differential equations. By taking the correlation time λ to be small we may again approximate $\varphi(\tau - \tau_1) \approx \lambda \delta(\tau - \tau_1)$, so that interactions are limited to a range $R \sim \lambda c$ and outgoing scattering trajectories are easily obtained by integration of the Lorentz force. This solution provides a reasonable qualitative description of the classical Bethe–Heitler process, which may be refined by numerical solution of the exact Lorentz force equations.

Initially (at time $\tau \to -\infty$), the target nucleus Z and incoming particle are widely separated. We set the nucleus at rest at the origin of the laboratory frame,

$$X_Z(\tau) = (ct_Z, \mathbf{x}_Z) = (c, \mathbf{0})\tau$$

and from some point x the line of observation

$$z = x - X_Z(\tau) = (ct, \mathbf{x}) - (c, \mathbf{0})\tau$$

satisfies

$$z^2 = (c(t - \tau), \mathbf{x})^2 = 0 \quad \longrightarrow \quad c(t - \tau) = R = |\mathbf{x}| \quad \longrightarrow \quad z = R\left(1, \hat{\mathbf{R}}\right),$$

where R is the scalar length defined in (4.9) as

$$-\frac{u \cdot z}{c} = -\frac{\dot{X}_Z \cdot z}{c} = -\frac{(c, \mathbf{0}) \cdot R\left(1, \hat{\mathbf{R}}\right)}{c} = R.$$

For the observation point we use the location of the incoming particle-1, approaching the nucleus on the trajectory

$$x = X_{in}(\tau) = (ct, \mathbf{x}) = u\tau + s = \dot{t}_{in}(c, v, 0, 0)\tau + (s_t, 0, s_y, 0),$$

where

$$u = \frac{d}{d\tau}(ct, x, y, z) \qquad \frac{dx}{d\tau} = \frac{dx}{dt}\dot{t}_{in} = v\dot{t}_{in} \qquad \dot{t}_{in} = \frac{dt}{d\tau} = \frac{1}{\sqrt{1 - \beta^2}}$$

and $\beta = v/c$. The scattering takes place in the plane $z = 0$ and since the nucleus is at the origin, we can write the spatial distance between the incoming particle and the target as

$$R(\tau) = |\mathbf{x}| = \sqrt{x^2 + y^2} = \sqrt{(v\dot{t}_{in}\tau)^2 + s_y^2}.$$

Putting $\lambda \approx R(\tau_1)$, the support of the fields is narrowly centered around the retarded time τ_1, so that τ_1 is determined from the causality conditions for the initial trajectories,

$$[X_{in}(\tau_1) - X_Z(\tau_1)]^2 = 0 \qquad X_{in}^0(\tau_1) - X_Z^0(\tau_1) > 0.$$

These equations have the solution

$$\tau_1 = \frac{1}{v\dot{t}_{in}(1 - \eta_v^2)}\left(\eta_v s_t + \sqrt{s_t^2 - s_y^2(1 - \eta_v^2)}\right) \xrightarrow[v \ll c]{} \frac{\sqrt{s_t^2 - s_y^2}}{v},$$

where we introduce the smooth parameter

$$\eta_v = \frac{1}{v}\left(1 - \frac{1}{\dot{t}_{in}}\right) \longrightarrow \begin{cases} 0, & v = 0 \\ 1, & v = c. \end{cases}$$

Notice that the 0-component s_t of the impact parameter must be positive in order for the interaction to take place. The location of the incoming particle at the time of interaction is now

$$\mathbf{x}(\tau_1) = R\,\hat{\mathbf{R}} \qquad\qquad t(\tau_1) = \dot{t}_{in}\tau_1 + s_t/c,$$

where

$$R = \frac{1}{1 - \eta_v^2}\left(\eta_v\sqrt{s_t^2 - s_y^2\left(1 - \eta_v^2\right)} + s_t\right) \xrightarrow[v \ll c]{} s_t$$

$$\hat{\mathbf{R}} = \frac{\left(\eta_v s_t + \sqrt{s_t^2 - s_y^2\left(1 - \eta_v^2\right)}, \left(1 - \eta_v^2\right)s_y, 0\right)}{s_t + \eta_v\sqrt{s_t^2 - s_y^2\left(1 - \eta_v^2\right)}} \xrightarrow[v \ll c]{} \left(\sqrt{1 - \frac{s_y^2}{s_t^2}}, \frac{s_y}{s_t}, 0\right).$$

Applying the Coulomb potential calculated in Section 4.1.1, the potential induced by the target nucleus in this approximation is

$$a^0(x, \tau) = \lambda\frac{Ze}{4\pi R}\delta(\tau - \tau_1) \qquad a^i = 0 \qquad a^5(x, \tau) = \frac{c_5}{c}a^0(x, \tau)$$

so that the nonzero field strengths

$$e^i = f^{0i} = \partial^0 a^i - \partial^i a^0 \qquad \epsilon^i = f^{5i} = \partial^5 a^i - \partial^i a^5 \qquad \epsilon^0 = \partial^5 a^0 - \partial^0 a^5$$

are

$$\mathbf{e} = -\nabla a^0 \qquad \boldsymbol{\epsilon} = -\nabla a^5 = \frac{c_5}{c}\mathbf{e} \qquad \epsilon^0 = \frac{\eta_{55}}{c_5}\left(1 + \frac{c_5^2}{c^2}\frac{1}{\dot{t}_{in}}\right)\partial_\tau a^0,$$

where we used

$$\frac{\partial}{\partial t}\delta(\tau - \tau_1) = \left(\frac{dt}{d\tau}\right)^{-1}_{t=t_{in}}\frac{\partial}{\partial\tau}\delta(\tau - \tau_1).$$

The nucleus and the incoming particle have opposite charge, so the Lorentz force

$$M\ddot{x}^0 = -\frac{e}{c}\left(f^{0i}\dot{x}_i + f^{05}\dot{x}_5\right) = -\frac{e}{c}\left(\mathbf{e}\cdot\dot{\mathbf{x}} - \eta_{55}c_5\epsilon^0\right)$$

$$M\ddot{\mathbf{x}} = -\frac{e}{c}\left(f^{k0}\dot{x}_0 + f^{k5}\dot{x}_5\right) = -\frac{e}{c}\left(ec\dot{i} - \eta_{55}c_5\boldsymbol{\epsilon}\right)$$

on the incoming particle becomes

$$\ddot{i} = \frac{\lambda Ze^2}{Mc^2}\left[\dot{\mathbf{x}}\cdot\nabla - \left(1 + \frac{c_5^2}{c^2}\frac{1}{\dot{t}_{in}}\right)\partial_\tau\right]\frac{\delta(\tau - \tau_1)}{4\pi R}$$

$$\ddot{\mathbf{x}} = \frac{\lambda Ze^2}{M}\left(\dot{i} - \eta_{55}\frac{c_5^2}{c^2}\right)\nabla\frac{\delta(\tau - \tau_1)}{4\pi R}.$$

The delta function enables immediate integration of the force equations as

$$
\dot{t}_f - \dot{t}_{in} = \frac{\lambda Z e^2}{M c^2} \int_{\tau_1-\lambda/2}^{\tau_1+\lambda/2} d\tau \left[\dot{\mathbf{x}} \cdot \nabla - \left(1 + \frac{c_5^2}{c^2}\frac{1}{\dot{t}_{in}}\right) \partial_\tau \right] \frac{\delta(\tau - \tau_1)}{4\pi R}
$$

$$
= \frac{\lambda Z e^2}{M c^2} \dot{\mathbf{x}}(\tau_1) \cdot \nabla \frac{1}{4\pi R}
$$

$$
= -\frac{\lambda}{M c^2} \frac{Z e^2}{4\pi R^2} \dot{\mathbf{x}}(\tau_1) \cdot \hat{\mathbf{R}} \tag{4.23}
$$

$$
\dot{\mathbf{x}}_f - \dot{\mathbf{x}}_{in} = \frac{\lambda Z e^2}{M} \int_{\tau_1-\lambda/2}^{\tau_1+\lambda/2} d\tau \left(\dot{t} - \eta_{55}\frac{c_5^2}{c^2} \right) \nabla \frac{\delta(\tau - \tau_1)}{4\pi R}
$$

$$
= -\frac{\lambda}{M} \frac{Z e^2}{4\pi R^2} \left(\dot{t}(\tau_1) - \eta_{55}\frac{c_5^2}{c^2} \right) \hat{\mathbf{R}}, \tag{4.24}
$$

where the velocities are evaluated at the interaction point as

$$
(\dot{t}, \dot{\mathbf{x}})(\tau_1) = \frac{1}{2}\left[(\dot{t}, \dot{\mathbf{x}})_f + (\dot{t}, \dot{\mathbf{x}})_{in} \right].
$$

We introduce the dimensionless parameter for Coulomb scattering

$$
g_e = \frac{\lambda}{M c}\frac{Z e^2}{4\pi R^2} = \frac{\lambda c}{R} \times \frac{Z e^2}{4\pi R}\frac{1}{M c^2} = \frac{\text{correlation length}}{\text{impact parameter}} \times \frac{\text{interaction energy}}{\text{mass energy}}
$$

which appears in (4.23) and (4.24) as the factor controlling the strength of the interaction. Writing

$$
\alpha_x = \frac{1}{2} g_e \hat{R}_x \qquad \alpha_y = \frac{1}{2} g_e \hat{R}_y
$$

we can expand the Lorentz force as components in the form

$$
\begin{bmatrix} 1 & \alpha_x & \alpha_y \\ \alpha_x & 1 & 0 \\ \alpha_y & 0 & 1 \end{bmatrix} \begin{bmatrix} c\dot{t}_f \\ \dot{x}_f \\ \dot{y}_f \end{bmatrix} = \begin{bmatrix} 1 & -\alpha_x & 0 \\ -\alpha_x & 1 & 0 \\ -\alpha_y & 0 & 0 \end{bmatrix} \begin{bmatrix} c\dot{t}_{in} \\ v\dot{t}_{in} \\ 0 \end{bmatrix} + 2\eta_{55}\frac{c_5^2}{c^2} \begin{bmatrix} 0 \\ \alpha_x \\ \alpha_y \end{bmatrix}
$$

and solve for the final velocity,

$$
\begin{bmatrix} c\dot{t}_f \\ \dot{x}_f \\ \dot{y}_f \end{bmatrix} = \frac{1}{1 - \frac{1}{4}g_e^2} \left\{ \begin{bmatrix} c\dot{t}_{in} \\ v\dot{t}_{in} \\ 0 \end{bmatrix} - g_e\dot{t}_{in} \begin{bmatrix} v\hat{R}_x \\ c\hat{R}_x \\ c\hat{R}_y \end{bmatrix} + \frac{1}{4}g_e^2\dot{t}_{in} \begin{bmatrix} c \\ v\left(\hat{R}_x^2 - \hat{R}_y^2\right) \\ 2v\hat{R}_x\hat{R}_y \end{bmatrix} \right\}, \tag{4.25}
$$

where we neglect $c_5^2/c^2 \ll \dot{t}_{in}$.

Before considering pair annihilation, we examine the low velocity and low interaction energy limit of this result. Taking

$$|\dot{\mathbf{x}}| = v \ll c \qquad \dot{t}_{in} \to 1 \qquad \eta_v \to 0 \qquad g_e \ll 1$$

the initial velocity reduces to

$$\dot{X}_{in}(\tau) \to (c, v, 0, 0),$$

the final velocity becomes

$$\dot{t}_f \approx \dot{t}_{in} \qquad \dot{\mathbf{x}}_f \approx \dot{\mathbf{x}} - g_e c \hat{\mathbf{R}} \qquad \hat{\mathbf{R}} = \left(\sqrt{1 - \frac{s_y^2}{s_t^2}}, \frac{s_y}{s_t}, 0 \right) \qquad R = s_t$$

and the scattering angle can be found as

$$\cos\theta = \frac{\dot{\mathbf{x}}_f \cdot \dot{\mathbf{x}}}{|\dot{\mathbf{x}}_f||\dot{\mathbf{x}}|} = \frac{\dot{\mathbf{x}}^2 - g_e c \hat{\mathbf{R}} \cdot \dot{\mathbf{x}}}{|\dot{\mathbf{x}}_f||\dot{\mathbf{x}}|} = \frac{v - g_e c \hat{R}_x}{|\dot{\mathbf{x}}_f|}.$$

If we also wish to impose the nonrelativistic condition for conservation of energy, we obtain a new constraint in the form

$$\dot{\mathbf{x}}^2 = v^2 = \dot{\mathbf{x}}_f^2 = \left[\dot{\mathbf{x}} - g_e c \hat{\mathbf{R}} \right]^2 \quad \Rightarrow \quad 2v\hat{R}_x = g_e c$$

in which case

$$\cos\theta = \frac{1}{|\dot{\mathbf{x}}_f|} \left[v - g_e c \hat{R}_x \right] = 1 - 2\hat{R}_x^2.$$

Now, using the definition of g_e we find

$$\cot\frac{\theta}{2} = \sqrt{\frac{1 + \cos\theta}{1 - \cos\theta}} = \frac{\hat{R}_y}{\hat{R}_x} = \frac{s_y}{s_t} \frac{2v}{g_e c} = \frac{2s_t}{\lambda v} \times \frac{4\pi M v^2 s_y}{Z e^2}$$

which recovers the Rutherford scattering formula if

$$\frac{2s_t}{\lambda v} = 1. \tag{4.26}$$

But for low energy we have $s_t = R(\tau_1)$ which we assumed to be comparable to λc. Since we cannot have $v \sim c$ in this low velocity case, (4.26) cannot be maintained. This result is unsurprising because the short-range potential cannot provide an adequate model of nonrelativistic Rutherford scattering.

Removing these restrictions and returning to the relativistic case, the condition for pair annihilation at τ_1 is that particle-1 scatters to negative energy, that is $\dot{t}_f < 0$ for some value of g_e which we call g_1. From (4.25),

$$\dot{t}_f = \dot{t}_{in} \frac{1 - g_1(v/c)\hat{R}_x + \frac{1}{4}g_1^2}{1 - \frac{1}{4}g_1^2}$$

and we see that for small values of g_1,

$$i_f \longrightarrow i_{in} \geq 1.$$

Since $v < c$ and $R_x < 1$, the numerator has discriminant

$$(vR_x/c)^2 - 1 < 0$$

and so is positive definite. The denominator becomes negative when

$$1 - \frac{1}{4}g_1^2 < 0 \quad \Rightarrow \quad g_1 = \frac{\text{correlation length}}{\text{impact parameter}} \times \frac{\text{interaction energy}}{\text{mass energy}} > 2$$

and since we take the correlation length λc approximately equal to the impact parameter R, the requirement for pair annihilation is

$$\frac{Ze^2}{4\pi R} > 2Mc^2$$

meaning that the interaction energy is greater than the mass energy of the annihilated particles. As g_1 approaches 2 from below i_f becomes very large. After g_1 passes this critical value, i_f decreases from large negative values, taking the limiting value

$$i_f \xrightarrow[g_1 \to \infty]{} -(i_{in} + 2) \qquad \Longrightarrow \qquad E_f = -(E_{in} + 2Mc^2)$$

so that the outgoing trajectory is timelike for all values of $g_1 > 2$.

Having found the condition for pair annihilation at time τ_1 we now consider the scattering at time τ_2, which we also treat as an incoming particle approaching a nucleus of opposite charge. Therefore we may apply the general expression (4.25). Particle-2 approaches a second nucleus along some trajectory $x_{in}^\mu(\tau)$ and emerges from the interaction along trajectory $x_f^\mu(\tau)$ with positive energy. The scattering and acceleration of particle-2 produces a radiation field which can be evaluated at some point of observation y^μ using the Liénard-Wiechert potential for an arbitrary trajectory.

The support of $\varphi(\tau - \tau_2)$ is narrowly centered on τ_2, and so the line of observation z^μ must be a lightlike vector, which we write as

$$z^\mu = y^\mu - x^\mu(\tau_2) = \rho \hat{\rho}^\mu \qquad \hat{\rho} = (1, \hat{\boldsymbol{\rho}}), \quad \hat{\boldsymbol{\rho}}^2 = 1.$$

We express the initial and final 4-velocities of the scattered particle as

$$\beta_{in} = \dot{x}_{in}/c \qquad\qquad \beta_f = \dot{x}_f/c$$

and define

$$\begin{aligned}
\Delta\beta &= \beta_f - \beta_{in} \\
\beta(\tau) &= \beta_{in} + \Delta\beta\,\theta(\tau - \tau_2) \\
\dot{\beta}(\tau) &= \Delta\beta\,\delta(\tau - \tau_2) \\
\beta(\tau_2) &= \bar{\beta} = \frac{1}{2}\left[\beta_f + \beta_{in}\right].
\end{aligned}$$

From (4.11) and (4.12) express the radiation fields produced by an arbitrary trajectory as

$$f_{rad}^{\mu\nu} = -e\varphi(\tau - \tau_2)\mathcal{F}^{\mu\nu}\left(z, \beta, \dot\beta\right) \qquad\qquad f_{rad}^{5\mu} = e\varphi(\tau - \tau_2)\mathcal{F}^{5\mu}\left(z, \beta, \dot\beta\right),$$

where

$$\mathcal{F}^{\mu\nu} = \left[\frac{\left(z \wedge \dot\beta\right)\rho - (z \wedge \beta)\left(\dot\beta \cdot z\right)}{4\pi c\rho^3}\right]^{\mu\nu} \qquad\qquad \mathcal{F}^{5\mu} = \frac{c_5}{c}\left[\frac{\left(\dot\beta \cdot z\right)z}{4\pi c\rho^3}\right]^{\mu}$$

and $\rho = -z \cdot \beta$ is the scalar distance from the scattered particle to the point of observation.

As pictured in Figure 4.1, the radiation emitted by the scattering of particle-2 is absorbed by the negative energy particle-1 arriving at y^μ. Using the Lorentz force equations we calculate the change in velocity of particle-1 caused by the incoming radiation. Since each term in the field strengths contains $\dot\beta(\tau) = \Delta\beta\,\delta(\tau - \tau_2)$ and $\varphi(0) = 1/2$, the change in velocity $\dot y^\mu(\tau_2)$ of particle-1 is

$$\Delta\dot y^\mu = \frac{e}{Mc}\int_{-\infty}^{\infty} d\tau\,\left[f_{rad}^{\mu\nu}\,\dot y_\nu + f_{rad}^{\mu 5}\,\dot y_5\right]$$

$$= \frac{e^2}{Mc}\int_{-\infty}^{\infty} d\tau\,\varphi(\tau - \tau_2)\left[-\mathcal{F}^{\mu\nu}\left(z, \beta, \dot\beta\right)\dot y_\nu - \eta_{55}c_5\mathcal{F}^{5\mu}\left(z, \beta, \dot\beta\right)\right]$$

$$= -\frac{e^2}{2Mc}\left[\mathcal{F}^{\mu\nu}\left(z, \bar\beta, \Delta\beta\right)\dot y_\nu + \eta_{55}c_5\mathcal{F}^{5\mu}\left(z, \bar\beta, \Delta\beta\right)\right] \qquad (4.27)$$

expressed in terms of the velocity change $\Delta\beta$ and average velocity $\bar\beta$. These are found from (4.25) to be

$$\Delta\beta = -\frac{g e}{1 - \frac{1}{4}g_e^2}t_{in}\begin{bmatrix}\beta\hat R_x \\ \hat R_x \\ \hat R_y\end{bmatrix} + \frac{\frac{1}{2}g_e^2}{1 - \frac{1}{4}g_e^2}t_{in}\begin{bmatrix}1 \\ \beta\hat R_x^2 \\ \beta\hat R_x\hat R_y\end{bmatrix}$$

$$\bar\beta = \frac{1}{1 - \frac{1}{4}g_e^2}t_{in}\begin{bmatrix}1 \\ \beta \\ 0\end{bmatrix} - \frac{\frac{1}{2}g e}{1 - \frac{1}{4}g_e^2}t_{in}\begin{bmatrix}\beta\hat R_x \\ \hat R_x \\ \hat R_y\end{bmatrix} + \frac{\frac{1}{4}g_e^2}{1 - \frac{1}{4}g_e^2}\beta t_{in}\begin{bmatrix}0 \\ -\hat R_y^2 \\ \hat R_x\hat R_y\end{bmatrix},$$

where now $\hat{\mathbf{R}}$ is the unit vector from the second nucleus to incoming particle-2 at the moment of scattering. Since particle-2 scatters at τ_2 to an $E > 0$ outgoing trajectory, we may take $v \ll c$ and so we set $g_e = g_2 < 1$ and $g_2^2 \approx 0$ for this interaction.

From (4.27) the Lorentz force acting on particle-1 at τ_2 can be written

$$\dot y_f^\mu + \frac{e^2}{2Mc}\left[\frac{(z \wedge \Delta\beta)\rho - (z \wedge \bar\beta)(\Delta\beta \cdot z)}{4\pi c\rho^3}\right]^{\mu\nu}\dot y_{vf}$$

$$= \dot y_{in}^\mu - \frac{e^2}{2Mc}\left[\frac{(z \wedge \Delta\beta)\rho - (z \wedge \bar\beta)(\Delta\beta \cdot z)}{4\pi c\rho^3}\right]^{\mu\nu}\dot y_\nu^{in}$$

neglecting the term $(c_5^2/c^2)\mathcal{F}^{5\mu}$. Making the simplifying choice $\hat{\mathbf{R}} \cdot \hat{\boldsymbol{\rho}} = 0$, we find

$$\bar{\beta} \cdot z = -\rho \dot{t}_{in}\left[1 - v\left(\hat{\rho}_x + \frac{1}{2}g_2\hat{R}_x\right)\right] \qquad \Delta\beta \cdot z = g_2\dot{t}_{in}v\rho\hat{R}_x,$$

where again we take $g_2^2 \approx 0$. Defining a second dimensionless factor for radiation

$$g_R = \frac{1}{2}\frac{e^2}{4\pi\rho}\frac{1}{Mc^2} = \frac{1}{2}\frac{\text{interaction energy}}{\text{mass energy}}$$

using

$$\left[(z \wedge \Delta\beta)\rho - (\Delta\beta \cdot z)(z \wedge \bar{\beta})\right] \cdot \dot{y} = z\,(\Delta\beta \cdot \dot{y})\rho - (z \cdot \dot{y})\Delta\beta\rho$$
$$- (\Delta\beta \cdot z)\left[(\bar{\beta} \cdot \dot{y})z - (z \cdot \dot{y})\bar{\beta}\right]$$

and now taking $\beta \approx 0$, the Lorentz force splits into the 0-component

$$\dot{y}_f^0 - g_2 g_R\hat{\mathbf{R}} \cdot \dot{\mathbf{y}}_f = \dot{y}_{in}^0 + g_2 g_R\hat{\mathbf{R}} \cdot \dot{\mathbf{y}}_{in}$$

and the space component

$$\dot{\mathbf{y}}_f - g_2 g_R\left[\left(\hat{\mathbf{R}} \cdot \dot{\mathbf{y}}_f\right)\hat{\boldsymbol{\rho}} + \left(\dot{y}_f^0 - \hat{\boldsymbol{\rho}} \cdot \dot{\mathbf{y}}_f\right)\hat{\mathbf{R}}\right] =$$
$$\dot{\mathbf{y}}_{in} + g_2 g_R\left[\left(\hat{\mathbf{R}} \cdot \dot{\mathbf{y}}_{in}\right)\hat{\boldsymbol{\rho}} + \left(\dot{y}_{in}^0 - \hat{\boldsymbol{\rho}} \cdot \dot{\mathbf{y}}_{in}\right)\hat{\mathbf{R}}\right].$$

We write the velocity of incoming negative energy particle-1 as

$$\dot{y}_{in}^0 < -1 \qquad \dot{\mathbf{y}}_{in} \cdot \hat{\boldsymbol{\rho}} = 0 \quad \Rightarrow \quad \dot{\mathbf{y}}_{in} = |\dot{\mathbf{y}}_{in}|\hat{\mathbf{R}}$$

and write the Lorentz force in components, with $g = g_2 g_R$, as

$$\begin{bmatrix} 1 & -g\hat{R}_x & -g\hat{R}_y \\ -g\hat{R}_x & 1 - g\hat{\rho}_x\hat{R}_x & -g\hat{\rho}_x\hat{R}_y \\ -g\hat{R}_y & -g\hat{\rho}_y\hat{R}_x & 1 - g\hat{\rho}_y\hat{R}_y \end{bmatrix}\begin{bmatrix} \dot{y}_f^0 \\ \dot{y}_{xf} \\ \dot{y}_{yf} \end{bmatrix} = \begin{bmatrix} 1 & 0 & 0 \\ g\hat{R}_x & 1 & 0 \\ g\hat{R}_y & 0 & 1 \end{bmatrix}\begin{bmatrix} \dot{y}_i^0 \\ \dot{y}_{xi} \\ \dot{y}_{yi} \end{bmatrix} + g\,|\dot{\mathbf{y}}_i|\begin{bmatrix} 1 \\ \hat{\rho}_x \\ \hat{\rho}_y \end{bmatrix}$$

so that the final velocity of particle-1 after absorbing the radiation is

$$\begin{bmatrix} \dot{y}_f^0 \\ \dot{\mathbf{y}}_f \end{bmatrix} = \frac{1}{1 - g^2}\begin{bmatrix} \dot{y}_{in}^0 \\ |\dot{\mathbf{y}}_{in}|\hat{\mathbf{R}} \end{bmatrix} + \frac{2g}{1 - g^2}\begin{bmatrix} |\dot{\mathbf{y}}_{in}| \\ \dot{y}_{in}^0\hat{\mathbf{R}} + |\dot{\mathbf{y}}_{in}|\hat{\boldsymbol{\rho}} \end{bmatrix}$$
$$+ \frac{g^2}{1 - g^2}\begin{bmatrix} \dot{y}_{in}^0 + |\dot{\mathbf{y}}_{in}| \\ 2\dot{y}_{in}^0\hat{\boldsymbol{\rho}} + 2\,|\dot{\mathbf{y}}_{in}|\hat{\mathbf{R}} \end{bmatrix} + \frac{g^3}{1 - g^2}\begin{bmatrix} |\dot{\mathbf{y}}_{in}| \\ |\dot{\mathbf{y}}_{in}|\hat{\boldsymbol{\rho}} \end{bmatrix}.$$

The 0-component is

$$\dot{y}_f^0 = \frac{1+g^2}{1-g^2}\dot{y}_{in}^0 + g\frac{2+g+g^2}{1-g^2}|\dot{\mathbf{y}}_{in}|$$

approximated at low velocity as

$$\dot{y}_f^0 \approx \frac{1+g^2}{1-g^2}\dot{y}_{in}^0 = -\alpha\dot{y}_{in}^0,$$

where

$$g^2 = \frac{\alpha+1}{\alpha-1} \quad \Rightarrow \quad \alpha = -\frac{1+g^2}{1-g^2}$$

is written so that $\alpha > 1$ for a positive energy timelike particle. The exact final velocity of the scattered particle is

$$\begin{bmatrix} \dot{y}_f^0 \\ \dot{\mathbf{y}}_f \end{bmatrix} = -\frac{\alpha-1}{2}\begin{bmatrix} \dot{y}_{in}^0 \\ |\dot{\mathbf{y}}_{in}|\,\hat{\mathbf{R}} \end{bmatrix} + \sqrt{\alpha^2-1}\begin{bmatrix} |\dot{\mathbf{y}}_{in}| \\ \dot{y}_{in}^0\hat{\mathbf{R}} + |\dot{\mathbf{y}}_{in}|\,\hat{\boldsymbol{\rho}} \end{bmatrix}$$

$$-\frac{\alpha+1}{2}\begin{bmatrix} \dot{y}_{in}^0 + |\dot{\mathbf{y}}_{in}| \\ 2\dot{y}_{in}^0\hat{\boldsymbol{\rho}} + 2|\dot{\mathbf{y}}_{in}|\,\hat{\mathbf{R}} \end{bmatrix} - \frac{\alpha+1}{2}\sqrt{\frac{\alpha+1}{\alpha-1}}\begin{bmatrix} |\dot{\mathbf{y}}_{in}| \\ |\dot{\mathbf{y}}_{in}|\,\hat{\boldsymbol{\rho}} \end{bmatrix}$$

with 0-component

$$\dot{y}_f^0 = -\alpha\dot{y}_{in}^0 - \frac{\alpha+1}{2}\left[1 + \frac{3-\alpha}{\sqrt{\alpha^2-1}}\right]|\dot{\mathbf{y}}_{in}|\,.$$

A pair creation event is observed at τ_2 for $\alpha > 1$ which requires that

$$g = g_2 g_R > 1 \qquad \longrightarrow \qquad \frac{e^2}{4\pi\rho} > \frac{2Mc^2}{g_2},$$

where $g_2 < 1$ and so the energy absorbed from the bremsstrahlung emitted from the scattering at τ_2 must be at least the total mass of the particle creation event observed in the laboratory. This provides a classical equivalent of the Bethe–Heitler mechanism in Stueckelberg–Horwitz–Piron electrodynamics.

4.7   PARTICLE MASS STABILIZATION

As we have seen, under the right circumstances a particle and an interacting pre-Maxwell field may exchange mass. In practical examples, such as pair creation and annihilation, the mass shift will be symmetric under evolution, so that the initial and final masses will be equal.

As another model of mass shift, consider an event propagating uniformly on-shell as

$$x(\tau) = u\tau = (u^0, \mathbf{u}) \qquad\qquad u^2 = -c^2$$

until it passes through a dense region of charged particles inducing

$$x(\tau) = u\tau + X(\tau),$$

where $X(\tau)$ is a small stochastic perturbation. If the typical distance scale between force centers is d then the perturbation will be roughly periodic with a characteristic period

$$\frac{d}{|\mathbf{u}|} = \frac{\text{a very short distance}}{\text{a moderate velocity}} = \text{a very short time,}$$

a fundamental frequency

$$\omega_0 = 2\pi \frac{|\mathbf{u}|}{d} = \text{very high frequency,}$$

and an amplitude on the order of

$$|X^\mu(\tau)| \sim \alpha d$$

for some macroscopic factor $\alpha < 1$. The perturbation can be represented in a Fourier series

$$X(\tau) = \mathrm{Re} \sum_n a_n \, e^{in\omega_0\tau} = \alpha d \, \mathrm{Re} \sum_n s_n^\mu \, e^{in\omega_0\tau}$$

with four-vector coefficients

$$a_n = \alpha d s_n = \alpha d \left(s_n^0, \mathbf{s}_n \right) = \alpha d \left(c s_n^t, \mathbf{s}_n \right),$$

where the s_n represent a normalized Fourier series ($s_0^\mu \sim 1$). The perturbed motion is of scale d, but the perturbed velocity

$$\dot{x}^\mu(\tau) = u^\mu + \dot{X}^\mu(\tau) = u^\mu + \alpha |\mathbf{u}| \, \mathrm{Re} \sum_n 2\pi n \, s_n^\mu \, i e^{in\omega_0\tau}$$

is of macroscopic scale $\alpha |\mathbf{u}|$. The unperturbed mass is $m = -M\dot{x}^2(\tau)/c^2 = M$ and the perturbed mass is

$$m = -\frac{M\dot{x}^2(\tau)}{c^2} = -\frac{M}{c^2}\left(u + \alpha |\mathbf{u}| \, \mathrm{Re} \sum_n 2\pi n \, s_n \, i e^{in\omega_0\tau} \right)^2$$

$$\simeq M \left(1 + 4\pi\alpha |\mathbf{u}| \, \mathrm{Re} \sum_n n \, s_n^t \, i e^{in\omega_0\tau} \right),$$

where we neglect terms in α^2. This kind of interaction may produce a macroscopic mass shift

$$m \longrightarrow m \left(1 + \frac{\Delta m}{m} \right) \qquad \frac{\Delta m}{m} = 4\pi\alpha |\mathbf{u}| \, \mathrm{Re} \sum_n n \, s_n^t \, i e^{in\omega_0\tau}$$

that remains significant after the interaction.

Two approaches have been suggested to explain why such mass shifts are not observed: one involving a self-interaction of the particle and its radiation field under mass shift, and the second a more general argument in statistical mechanics.

4.7.1   SELF-INTERACTION

We consider an arbitrarily moving event $X^\mu(\tau)$ at the origin of a co-moving frame so that

$$X(\tau) = (ct(\tau), \mathbf{0}) \qquad \dot{X}(\tau) = (c\dot{t}(\tau), \mathbf{0})$$

and a change in mass $m = -M\dot{X}^2/c^2 = M\dot{t}^2$ can only result from a change in energy through acceleration of t. We say that the event is on-shell if $\dot{t} = 1$. The Green's function permits us to compute the field at some point x induced by the evolving event. If the motion at time τ produces an observable field at time $\tau^* > \tau$ at some point $x = X(\tau^*)$ along the trajectory of the event itself, then the event will experience a self-force. Because $G_{Maxwell} = 0$ on the event's timelike trajectory, only a contribution from $G_{Correlation}$ can produce such a self-interaction, and, as seen from (3.24), only if $\eta_{55} = +1$.

We approximate $\varphi(\tau' - s) = \lambda\delta(\tau' - s)$ as in Section 4.1.2, introduce the function $g(s)$ to express terms of the type

$$c^2 g(s) = -\left((X(\tau^*) - X(s))^2 + c_5^2(\tau^* - s)^2 \right) = c^2 \left((t(\tau^*) - t(s))^2 - \frac{c_5^2}{c^2}(\tau^* - s)^2 \right)$$

and write

$$a^\alpha \left(X(\tau^*), \tau^* \right) = \frac{\lambda e c_5}{2\pi^2 c^3} \int ds\, \dot{X}^\alpha(s) \left(\frac{1}{2} \frac{\theta(g(s))}{(g(s))^{3/2}} - \frac{\delta(g(s))}{(g(s))^{1/2}} \right) \theta^{ret}$$

for the self-field experienced by the event. We designate the two terms as

$$a^\alpha \left(X(\tau^*), \tau^* \right) = a_\theta^\alpha + a_\delta^\alpha.$$

For an event evolving uniformly on-shell we have

$$t(\tau^*) = \tau^* \qquad g(s) = \left(1 - \frac{c_5^2}{c^2} \right)(\tau^* - s)^2$$

and using identity (4.7) are led to

$$a \left(X(\tau^*), \tau^* \right) = \frac{\lambda e c_5}{2\pi^2 c^3} (c, \mathbf{0}, c_5) \int ds\, \theta(\tau^* - s)$$

$$\left(\frac{1}{2} \frac{\theta\left(\left(1 - \frac{c_5^2}{c^2}\right)(\tau^* - s)^2 \right)}{\left(\left(1 - \frac{c_5^2}{c^2}\right)(\tau^* - s)^2 \right)^{3/2}} - \frac{\delta\left(\left(1 - \frac{c_5^2}{c^2}\right)(\tau^* - s)^2 \right)}{\left(\left(1 - \frac{c_5^2}{c^2}\right)(\tau^* - s)^2 \right)^{1/2}} \right)$$

$$= \frac{\lambda e c_5 (c, \mathbf{0}, c_5)}{2\pi^2 c^3 \left(1 - \frac{c_5^2}{c^2} \right)^{3/2}} \int_{-\infty}^{\tau^*} ds \left(\frac{1}{2} \frac{1}{(\tau^* - s)^3} - \frac{\delta(\tau^* - s)\,\theta(\tau^* - s)}{\left| (\tau^* - s)^2 \right|} \right).$$

Since

$$\int_{-\infty}^{\tau^*} ds \, \frac{1}{(\tau^* - s)^3} = \frac{1}{2(\tau^* - s)^2}\bigg|_{-\infty}^{\tau^*} = \lim_{s \to \tau^*} \frac{1}{2(\tau^* - s)^2}$$

and

$$\int_{-\infty}^{\tau^*} ds \, \frac{\delta(\tau^* - s)\theta(\tau^* - s)}{(\tau^* - s)^2} = \lim_{s \to \tau^*} \frac{\theta(\tau^* - s)}{(\tau^* - s)^2} = \lim_{s \to \tau^*} \frac{\frac{1}{2}}{(\tau^* - s)^2}$$

we find that for uniform on-shell motion

$$a\left(X(\tau^*), \tau^*\right) = \frac{\lambda e c_5}{2\pi^2 c^3}(c, 0, c_5) \lim_{s \to \tau^*}\left(\frac{1}{2(\tau^* - s)^2} - \frac{\frac{1}{2}}{(\tau^* - s)^2}\right) = 0$$

the self-force vanishes.

In general, because $\dot{X}^i = 0$ and $a^\alpha(X(\tau^*), \tau^*)$ does not depend on X^i, we have

$$a^i = 0 \qquad \partial_i a^0 = \partial_i a^5 = 0 \qquad \Rightarrow \qquad f^{\mu\nu} = f^{5i} = 0$$

and so the field reduces to

$$f^{50} = \partial^5 a^0 - \partial^0 a^5 = \frac{1}{c_5}\partial_{\tau^*}a^0 + \frac{1}{c}\partial_t a^5,$$

where the partial derivative ∂_{τ^*} only acts on the explicit variable (not on $t(\tau^*)$ or θ^{ret}). Similarly, the velocity $\dot{X}^\alpha(s)$ is constant with respect to ∂_{τ^*}.

Inserting the potential we find

$$\partial^5 a_\theta^0 - \partial^0 a_\theta^5 = \frac{3\lambda e c_5}{4\pi^2 c^3}\frac{c_5}{c}\int ds \, \frac{\theta\left((t(\tau^*) - t(s))^2 - \frac{c_5^2}{c^2}(\tau^* - s)^2\right)}{\left[(t(\tau^*) - t(s))^2 - \frac{c_5^2}{c^2}(\tau^* - s)^2\right]^{5/2}}\theta^{ret}\,\Delta(\tau^*, s)$$

$$-\frac{\lambda e c_5}{2\pi^2 c^3}\frac{c_5}{c}\int ds \, \frac{\delta\left((t(\tau^*) - t(s))^2 - \frac{c_5^2}{c^2}(\tau^* - s)^2\right)}{\left[(t(\tau^*) - t(s))^2 - \frac{c_5^2}{c^2}(\tau^* - s)^2\right]^{3/2}}\theta^{ret}\,\Delta(\tau^*, s),$$

where

$$\Delta(\tau^*, s) = \dot{t}(s)(\tau^* - s) - \left(t(\tau^*) - t(s)\right)$$

characterizes the energy acceleration in the rest frame, which will be associated with mass shift. Similarly, the derivatives of a_8 produce

$$\partial^5 a_8^0 - \partial^0 a_8^5 = -\frac{\lambda e c_5}{2\pi^2 c^3}\frac{c_5}{c}\int ds \frac{\delta\left((t\,(\tau^*) - t\,(s))^2 - \frac{c_5^2}{c^2}(\tau^* - s)^2\right)}{\left((t\,(\tau^*) - t\,(s))^2 - \frac{c_5^2}{c^2}(\tau^* - s)^2\right)^{3/2}}\theta^{ret}\,\Delta\,(\tau^*,s)$$

$$-\frac{\lambda e c_5}{2\pi^2 c^3}\frac{c_5}{c}\int ds \frac{2\delta'\left((t\,(\tau^*) - t\,(s))^2 - \frac{c_5^2}{c^2}(\tau^* - s)^2\right)}{\left((t\,(\tau^*) - t\,(s))^2 - \frac{c_5^2}{c^2}(\tau^* - s)^2\right)^{1/2}}\theta^{ret}\,\Delta\,(\tau^*,s)$$

and combining terms we find

$$f^{50} = f_\theta^{50} + f_8^{50} + f_{8'}^{50},$$

where

$$f_\theta^{50} = \frac{3\lambda e}{4\pi^2}\frac{c_5^2}{c^4}\int ds \frac{\theta\left((t\,(\tau^*) - t\,(s))^2 - \frac{c_5^2}{c^2}(\tau^* - s)^2\right)}{\left[(t\,(\tau^*) - t\,(s))^2 - \frac{c_5^2}{c^2}(\tau^* - s)^2\right]^{5/2}}\theta^{ret}\,\Delta\,(\tau^*,s) \qquad (4.28)$$

$$f_8^{50} = -\frac{\lambda e}{\pi^2}\frac{c_5^2}{c^4}\int ds \frac{\delta\left((t\,(\tau^*) - t\,(s))^2 - \frac{c_5^2}{c^2}(\tau^* - s)^2\right)}{\left[(t\,(\tau^*) - t\,(s))^2 - \frac{c_5^2}{c^2}(\tau^* - s)^2\right]^{3/2}}\theta^{ret}\,\Delta\,(\tau^*,s) \qquad (4.29)$$

$$f_{8'}^{50} = -\frac{\lambda e}{\pi^2}\frac{c_5^2}{c^4}\int ds \frac{\delta'\left((t\,(\tau^*) - t\,(s))^2 - \frac{c_5^2}{c^2}(\tau^* - s)^2\right)}{\left((t\,(\tau^*) - t\,(s))^2 - \frac{c_5^2}{c^2}(\tau^* - s)^2\right)^{1/2}}\theta^{ret}\,\Delta\,(\tau^*,s). \qquad (4.30)$$

Notice that if the particle remains at constant velocity (in any uniform frame), then

$$x^0\,(\tau) = u^0\tau \qquad \longrightarrow \qquad \Delta\,(\tau^*,s) = \frac{u^0}{c}(\tau^* - s) - \left(\frac{u^0}{c}\tau^* - \frac{u^0}{c}s\right) = 0$$

and so the self-force f^{50} vanishes. For any smooth $t\,(\tau)$, we may approximate

$$t\,(\tau^*) - t\,(s) = t\,(s) + \dot{t}(s)(\tau^* - s) + \frac{1}{2}\ddot{t}(s)(\tau^* - s)^2 + o\left((\tau^* - s)^3\right) - t\,(s)$$

$$= \dot{t}(s)(\tau^* - s) + \frac{1}{2}\ddot{t}(s)(\tau^* - s)^2 + o\left((\tau^* - s)^3\right)$$

so the function

$$\Delta\left(\tau^*,s\right) = \dot{t}(s)(\tau^* - s) - \left(t\left(\tau^*\right) - t\left(s\right)\right) = -\frac{1}{2}\ddot{t}(s)(\tau^* - s)^2 + o\left((\tau^* - s)^3\right)$$

is nonzero only when the time coordinate accelerates in the rest frame, equivalent to a shift in the particle mass.

As a first-order example, we consider a small, sudden jump in mass at $\tau = 0$ characterized by

$$t\left(\tau\right) = \begin{cases} \tau & , \quad \tau < 0 \\ (1+\beta)\,\tau & , \quad \tau > 0 \end{cases} \qquad \Rightarrow \qquad \dot{t}\left(\tau\right) = \begin{cases} 1 & , \quad \tau < 0 \\ 1+\beta & , \quad \tau > 0 \end{cases}$$

and calculate the self-interaction. Since θ^{ret} enforces $t(\tau^*) > t(s)$, it follows that

$$\tau^* < 0 \quad \Rightarrow \quad s < 0 \quad \Rightarrow \quad \dot{t}(\tau^*) = \dot{t}(s) = 1 \quad \Rightarrow \quad \Delta(\tau^*,s) = 0.$$

Similarly,

$$\tau^* > 0 \ \text{ and } \ s > 0 \quad \Rightarrow \quad \dot{t}(\tau^*) = \dot{t}(s) = 1 + \beta \quad \Rightarrow \quad \Delta(\tau^*,s) = 0.$$

But when $\tau^* > 0$ and $s < 0$,

$$\Delta(\tau^*,s) \doteq \dot{t}(s)(\tau^* - s) - \left(t\left(\tau^*\right) - t\left(s\right)\right) = (\tau^* - s) - \left[(1 + \beta)\left(\tau^*\right) - s\right] = -\beta\tau^*$$

and f^{50} can be found from the contributions (4.28)–(4.30). Writing

$$g\left(s\right) = \left(t\left(\tau^*\right) - t\left(s\right)\right)^2 - \frac{c_5^2}{c^2}(\tau^* - s)^2 = \left((1 + \beta)\,\tau^* - s\right)^2 - \frac{c_5^2}{c^2}(\tau^* - s)^2$$

and solving for $g(s^*) = 0$, we find

$$s^* = \left(1 + \frac{\beta}{1 - \dfrac{c_5}{c}}\right)\tau^* > \tau^*$$

so that $g(s) > 0$ in the region of interest $s < 0 < \tau^*$ and there will be no contribution from the terms (4.29) or (4.30). Thus,

$$f^{50} = f_\theta^{50} = (-\beta\tau^*)\frac{3\lambda e}{4\pi^2}\frac{c_5^2}{c^4}\int_{-\infty}^{0} ds\, \frac{1}{\left[(t\left(\tau^*\right) - t\left(s\right))^2 - \dfrac{c_5^2}{c^2}(\tau^* - s)^2\right]^{5/2}}$$

$$= (-\beta\tau^*)\frac{3\lambda e}{4\pi^2}\frac{c_5^2}{c^4}\int_{-\infty}^{0} ds\, \frac{1}{\left[((1 + \beta)\,\tau^* - s)^2 - \dfrac{c_5^2}{c^2}(\tau^* - s)^2\right]^{5/2}}.$$

Shifting the integration variable as $x = \tau^* - s$ the integral becomes

$$\int_{-\infty}^{0} ds \, \frac{1}{\left[((1+\beta)\,\tau^* - s)^2 - \frac{c_5^2}{c^2}(\tau^* - s)^2 \right]^{5/2}} = -\int_{\infty}^{\tau^*} \frac{dx}{(Cx^2 + Bx + A)^{5/2}},$$

where

$$C = 1 - \frac{c_5^2}{c^2} \qquad B = 2\beta\tau^* \qquad A = (\beta\tau^*)^2$$

which can be evaluated using the well-known form [3]

$$\int \frac{dx}{(Cx^2 + Bx + A)^{5/2}} = \frac{2(2Cx + B)}{3q\sqrt{Cx^2 + Bx + A}} \left(\frac{1}{Cx^2 + Bx + A} + \frac{8C}{q} \right),$$

where $q = 4AC - B^2$. We finally find the field strength in the form

$$f^{50} = \frac{\lambda e}{4\pi^2} \frac{1}{c_5^2 (\beta\tau^*)^3} \, Q\left(\beta, \frac{c_5^2}{c^2} \right),$$

where $Q\left(\beta, \frac{c_5^2}{c^2} \right)$ is the positive, dimensionless factor

$$Q\left(\beta, \frac{c_5^2}{c^2} \right) = \left[2\left(1 - \frac{c_5^2}{c^2} \right)^{3/2} \left(1 - \frac{\left(1 - \frac{c_5^2}{c^2} \right)^{1/2} \left(1 + \frac{\beta}{\left(1 - \frac{c_5^2}{c^2} \right)} \right)}{\left[1 + \frac{2\beta}{1 - \frac{c_5^2}{c^2}} + \frac{\beta^2}{1 - \frac{c_5^2}{c^2}} \right]^{1/2}} \right) \right.$$

$$\left. + \frac{\beta^2 \frac{c_5^2}{c^2} \left(1 + \frac{c_5^2}{c^2} \frac{\beta}{1 - \frac{c_5^2}{c^2}} \right)}{\left(1 - \frac{c_5^2}{c^2} \right)^{1/2} \left[1 + \frac{2\beta}{1 - \frac{c_5^2}{c^2}} + \frac{\beta^2}{1 - \frac{c_5^2}{c^2}} \right]^{3/2}} \right]$$

which is seen to be finite for $c_5 < c$, with

$$Q\left(\beta, \frac{c_5^2}{c^2}\right) \xrightarrow[c_5 \to 0]{} 2\left(1 - \frac{1+\beta}{[1 + 2\beta + \beta^2]^{1/2}}\right) = 0.$$

Since $f^{\mu\nu} = 0$, the Lorentz force induced by this field strength is then

$$M\ddot{x}^{\mu} = ef^{\mu\alpha}\dot{x}_{\alpha} = ef^{\mu 5}\dot{x}_5 = -ef^{5\mu}\dot{x}_5 = -\eta_{55}ef^{5\mu}\dot{x}^5 = -ef^{5\mu}c_5$$

and since $f^{5i} = 0$

$$M\ddot{x}^i = 0$$

$$M\ddot{x}^0 = -c_5 ef^{50} = \begin{cases} 0 & , \quad \tau^* < 0 \\ -\dfrac{\lambda e^2}{4\pi^2} \dfrac{1}{c_5 \, (\beta\tau^*)^3} \, Q\left(\beta, \dfrac{c_5^2}{c^2}\right) & , \quad \tau^* > 0 \end{cases}$$

which causes the 0-coordinate to decelerate. When the event returns to on-shell propagation the function $\Delta(\tau^*, s)$ and field strength f^{50} again vanish. The mass decay can also be seen in the Lorentz force for the mass

$$\frac{d}{d\tau}\left(-\frac{1}{2}M\dot{x}^2\right) = ef^{5\mu}\dot{x}_{\mu} = ef^{50}\dot{x} = -ecf^{50}\dot{t} = -\frac{\lambda e^2}{4\pi^2}\frac{c}{c_5^2 \, (\beta\tau^*)^3} \, Q\left(\beta, \frac{c_5^2}{c^2}\right)\dot{t}.$$

We notice that if $\beta < 0$ then f^{50} changes sign so that the self-interaction results in damping or anti-damping to push the trajectory toward on-shell behavior. Although this model is approximate, it seems to indicate that the self-interaction of the event with the field generated by its mass shift will restore the event to on-shell propagation.

4.7.2   STATISTICAL MECHANICS

In Section 3.4 we saw that a particle, as observed through its electromagnetic current, can be interpreted as a weighted ensemble of events $\varphi(s)x^{\mu}(\tau + s)$ selected from a neighborhood of event $x^{\mu}(\tau)$ (along a single timelike trajectory) determined by $\varphi(s)$. Here we model a particle as an ensemble $x_i^{\mu}(\tau)$ of N mutually interacting event trajectories given at a single τ. Constructing the canonical and grand canonical ensembles without an *a priori* constraint on the total mass of the system, the total mass of the particle is determined by a chemical potential. Under perturbation, such as collisions for which the final asymptotic mass of an elementary event is not constrained by the basic theory, the particle returns to its equilibrium mass value. Here we provide here a brief summary of the full model given in [12, 13].

As described in Section 2.5, we first construct a canonical ensemble by extracting a small subensemble Γ_s (the particle system) from its environment Γ_b (the bath ensemble). Summing over all possible partitions of energy and mass parameter between the particle and bath

$$\Gamma(\kappa, E) = \int d\Omega_b d\Omega_s d\kappa_b d\kappa_s \delta(K_b - \kappa_b)\delta(K_s - \kappa_s)\delta(E_s + E_b - E)\delta(\kappa_s + \kappa_b - \kappa)$$
$$= \int d\kappa' dE' \Gamma_b(\kappa - \kappa', E - E')\Gamma_s(\kappa', E')$$

in which both mass and energy may be exchanged. We suppose that the integrand has a maximum over both variables κ', E', providing an equilibrium point for the system. By analyzing the partial derivatives, it can be shown that no saddle point configuration is possible in the neighborhood of the maximum. The conditions for equilibrium can then be written

$$\frac{1}{\Gamma_b(\kappa - \kappa', E - E')}\frac{\partial \Gamma_b}{\partial E}(\kappa - \kappa', E - E')|_{\max} = \frac{1}{\Gamma_s(\kappa', E')}\frac{\partial \Gamma_s}{\partial E}(\kappa', E')|_{\max} \equiv \frac{1}{T}$$

and

$$\frac{1}{\Gamma_b(\kappa - \kappa', E - E')}\frac{\partial \Gamma_b}{\partial \kappa}(\kappa - \kappa', E - E')|_{\max} = \frac{1}{\Gamma_s(\kappa', E')}\frac{\partial \Gamma_s}{\partial \kappa}(\kappa', E)|_{\max} \equiv \frac{1}{T_\kappa},$$

defining temperature in the usual way, and a new effective "mass temperature" T_κ. Writing

$$S_b(\kappa, E) = \ln \Gamma_b(\kappa, E) \qquad\qquad S_s(\kappa, E) = \ln \Gamma_s(\kappa, E)$$

it follows that at maximum

$$\frac{\partial S_b}{\partial E} = \frac{\partial S_s}{\partial E} = \frac{1}{T} \qquad\qquad \frac{\partial S_b}{\partial \kappa} = \frac{\partial S_s}{\partial \kappa} = \frac{1}{T_\kappa}.$$

By additivity of entropy, the total entropy of the system is independent of κ', E' in the neighborhood of the maximum, and for κ' and E' small compared to κ and E,

$$\Gamma_b(\kappa - \kappa', E - E') = e^{S_b(\kappa-\kappa',E-E')} \cong e^{S_b(\kappa,E)-\kappa'\frac{\partial S_b}{\partial \kappa}-E'\frac{\partial S_b}{\partial E}} = e^{S_b(\kappa,E)}e^{-\frac{\kappa'}{T_\kappa}}e^{-\frac{E'}{T}}$$

in this neighborhood. Then

$$\Gamma(\kappa, E) = \int d\kappa' dE' \Gamma_s(\kappa', E')e^{S_b(\kappa,E)}e^{-\frac{\kappa'}{T_\kappa}}e^{-\frac{E'}{T}} = e^{S_b(\kappa,E)}\int d\Omega_s e^{-\frac{K_s}{T_\kappa}}e^{-\frac{E_s}{T}}$$

leading to the partition function

$$Q_N(T_\kappa, T) = \int d\Omega e^{-\frac{K}{T_\kappa}}e^{-\frac{E}{T}},$$

where the overall factor $S_b(\kappa, E)$ cancels out in any computation of average values. The Helmholtz free energy A is defined through

$$Q_N(T_\kappa, T) = e^{-A(T_\kappa,T)/T} \qquad\qquad \int d\Omega e^{-K/T_\kappa}e^{(A-E)/T} = 1$$

from which it follows that

$$A = \langle E \rangle + T\frac{\partial A}{\partial T} = \langle E \rangle - TS \qquad\qquad S = -\frac{\partial A}{\partial T}$$

and

$$\langle K \rangle = -\frac{T_\kappa^2}{T}\frac{\partial A}{\partial T_\kappa}.$$

Under the canonical distribution, corresponding to an equilibrium of both heat and mass, without exchange of particles with the bath, we therefore obtain a mean value for $\langle K \rangle$, the effective center-of-mass mass of the subensemble, which is determined by T_κ and T.

Computing the fluctuations in energy, one finds

$$\langle (E - \langle E \rangle)^2 \rangle = T^2\frac{\partial \langle E \rangle}{\partial T} \qquad\qquad \langle (K - \langle K \rangle)^2 \rangle = T_\kappa{}^2\frac{\partial \langle K \rangle}{\partial T_\kappa}$$

showing that the mean mass rises with the mass temperature. (Since K is proportional to a negative mass in this metric, $-T_\kappa$ is a positive number, to be identified with a "mass temperature.")

Repeating the above for the grand canonical ensemble, in which the system (particle) ensemble may exchange events and volume with the bath, one decomposes the full microcanonical in terms of its canonical subsets

$$Q_N(V, T, T_\kappa) = \sum_{N_s=0}^{N} \int d\Omega_s e^{-K_s/T_\kappa}e^{-E_s/T} Q_{N-N_s}(V - V_s, T, T_\kappa),$$

where

$$Q_{N-N_s}(V - V_s, T, T_\kappa) = \int d\Omega_b e^{-K_b/T_\kappa}e^{-E_b/T}$$

$K_b = K - K_s$ and $E_b = E - E_s$. Making the usual identifications

$$\frac{\partial A}{\partial V} = -P \qquad\qquad \frac{\partial A}{\partial N} = \mu$$

and defining the new mass chemical potential

$$\frac{\partial A}{\partial K} = -\mu_\kappa$$

leads to the grand partition function

$$\mathcal{Q}(V, T, T_\kappa) = e^{VP/T} = \sum_{N_s=0}^{N} z^{N_s} Q_{N_s}(T, K_s, E_s),$$

where

$$Q_{N_s}(T, K_s, E_s) = \int d\Omega_s \zeta^{K_s}e^{-E_s/T} \qquad\qquad z = e^{\mu/T} \qquad\qquad \zeta = e^{-\hat\mu_\kappa/T}.$$

It follows that

$$\langle N \rangle = z \frac{\partial}{\partial z} \ln Q \qquad\qquad \langle K \rangle = \zeta \frac{\partial}{\partial \zeta} \ln Q.$$

Modifying the Helmholtz free energy for the grand canonical ensemble,

$$A = \langle N \rangle\, T \ln z + \langle K \rangle\, T \ln \zeta - T \ln Q$$

leads to

$$Q = e^{-A/T} z^{<N>} \zeta^{<K>}.$$

It follows that the internal energy is

$$U \equiv \langle E \rangle = A - \langle N \rangle \mu + \left(\mu_\kappa + \frac{T}{T_\kappa} \right) \langle K \rangle + T \ln Q + T^2 \frac{\partial}{\partial T} \ln Q$$

and using the thermodynamic relation

$$U = A + TS$$

one finds

$$S = \frac{\partial}{\partial T} (T \ln Q) + \left(\frac{\mu_\kappa}{T} + \frac{1}{T_\kappa} \right) \langle K \rangle - \frac{\mu}{T} \langle N \rangle.$$

Finally, the Maxwell relations are

$$S = - \left(\frac{\partial A}{\partial T} \right) \Bigg|_{V, \langle N \rangle, \langle K \rangle} \qquad\qquad P = - \left(\frac{\partial A}{\partial V} \right) \Bigg|_{<N>,<K>,T}$$

$$\frac{\partial A}{\partial \langle N \rangle} = \mu \qquad\qquad \frac{\partial A}{\partial \langle K \rangle} = - \left(\mu_\kappa + \frac{T}{T_\kappa} \right).$$

At the critical point in $\langle K \rangle$

$$\frac{\partial A}{\partial \langle K \rangle} = 0 \qquad \longrightarrow \qquad \frac{T}{T_\kappa} = -\mu_\kappa \qquad\qquad (4.31)$$

and so μ_κ is positive since T_κ is negative.

The particle in this model is a statistical ensemble which has both an equilibrium energy and an equilibrium mass, controlled by the temperature and chemical potentials, thus assuring asymptotic states with the correct mass. The thermodynamic properties of this system, involve the maximization of the integrand in the microcanonical ensemble, where both the energy and the mass are parameters of the distribution. A critical point in the free energy is made available by the interplay of the equilibrium requirements of the canonical ensemble (where the total mass of the system is considered variable) as for the energy, and the equilibrium requirements of the grand canonical ensemble (where a chemical potential arises for the particle number). The particle mass is controlled by a chemical potential, so that asymptotic variations in the mass can be restored to a given value by relaxation to satisfy the equilibrium conditions.

4.8   SPEEDS OF LIGHT AND THE MAXWELL LIMIT

As discussed in Section 3.7, concatenation—integration of the pre-Maxwell field equations over the evolution parameter τ—extracts from the microscopic event interactions the massless modes in Maxwell electrodynamics, expressing a certain equilibrium limit when mass exchange settles to zero. In this picture, the microscopic dynamics approach an equilibrium state because the boundary conditions hold pointwise in x as $\tau \to \infty$, asymptotically eliminating interactions that cannot be described in Maxwell theory. The Maxwell-type description recovered by concatenating the microscopic dynamics may thus be understood as a self-consistent summary constructed *a posteriori* from the complete worldlines.

We have assumed that $0 \leq c_5 < c$ and we must check that SHP theory remains finite as $c_5 \to 0$. First we notice that c_5 appears explicitly three times in the pre-Maxwell equations (3.20)

$$\partial_\nu f^{\mu\nu} - \frac{1}{c_5} \partial_\tau f^{5\mu} = \frac{e}{c} j_\varphi^\mu \qquad \partial_\mu f^{5\mu} = \frac{e}{c} j_\varphi^5 = \frac{c_5}{c} e\rho_\varphi$$

$$\partial_\mu f_{\nu\rho} + \partial_\nu f_{\rho\mu} + \partial_\rho f_{\mu\nu} = 0 \qquad \partial_\nu f_{5\mu} - \partial_\mu f_{5\nu} + \frac{1}{c_5} \partial_\tau f_{\mu\nu} = 0$$

twice in the form $\frac{1}{c_5} \partial_\tau$ and once multiplying the event density ρ_φ. The derivative term poses no problem in the homogeneous pre-Maxwell equation, which is satisfied identically for fields derived from potentials. Specifically, the fields $f_{5\mu}$ contain terms of the type $\partial_5 a_\mu = \frac{1}{c_5} \partial_\tau a_\mu$ that cancel the explicit τ-derivative of $f_{\mu\nu}$, evaluated before passing to the limit $c_5 \to 0$. However, the homogeneous equation does impose a new condition through

$$c_5 \left(\partial_\nu f_{5\mu} - \partial_\mu f_{5\nu} \right) + \partial_\tau f_{\mu\nu} = 0 \quad \xrightarrow[c_5 \to 0]{} \quad \partial_\tau f_{\mu\nu} = 0$$

requiring that the field strength $f^{\mu\nu}$ become τ-independent in this limit. For the fields derived in Section 4.2 this condition is violated by the multiplicative factor $\varphi(\tau - \tau_R)$ unless we simultaneously require $c_5 \to 0 \Rightarrow \lambda \sim 1/c_5 \to \infty$, in which case $\varphi(x,\tau) \to 1/2\xi = 1$, using (3.12) for ξ. This requirement effectively spreads the event current j_φ^α uniformly along the particle worldline, recovering the τ-independent particle current

$$j_\varphi^\mu (x,\tau) = \int ds\, \varphi (\tau - s)\, j^\mu (x,s) \quad \longrightarrow \quad \int ds\, 1 \cdot j^\mu (x,s) = J^\mu(x)$$

$$j_\varphi^5 (x,\tau) = \int ds\, \varphi (\tau - s)\, j^5 (x,s) \quad \longrightarrow \quad \int ds\, j^5 (x,s)$$

$$\partial_\mu j_\varphi^\mu (x,\tau) + \frac{1}{c_5} \partial_\tau j_\varphi^5 (x,\tau) \quad \longrightarrow \quad \partial_\mu J^\mu (x) = 0$$

associated with Maxwell theory. Generally, because the τ-dependence of the potentials and fields is contained in φ, the condition $\lambda \to \infty$ eliminates all the terms in the pre-Maxwell equations containing ∂_τ. Similarly, the photon mass $m_\gamma \sim \hbar/\xi\lambda c^2$ must vanish.

We saw that $f^{5\mu}$ is generally proportional to c_5 for fields of the Liénard–Wiechert type. Therefore, we can write the inhomogeneous pre-Maxwell equations in the finite form

$$\partial_\nu \, f^{\mu\nu} = \frac{e}{c} \, j_\varphi^\mu \qquad\qquad \partial_\mu \left(\frac{1}{c_5} f^{5\mu} \right) = \frac{e}{c} \, \rho_\varphi,$$

where we see that $f^{5\mu}$ decouples from the field $f^{\mu\nu}$ that now satisfies Maxwell's equations.

To find the limiting form of the electromagnetic interactions, we consider an arbitrary event $X^\mu \, (\tau)$, which induces the current

$$j_\varphi^\alpha \, (x, \tau) = c \int ds \; \varphi \, (\tau - s) \, \dot{X}^\alpha \, (s) \, \delta^4 \, [x - X \, (s)].$$

From the field strengths found in Section 4.2 the Lorentz force on a test event moving in the field induced by this current can be written

$$M\ddot{x}^\mu = \frac{e}{c} \left[f^\mu_{\ \nu}(x, \tau)\dot{x}^\nu + f^{5\mu}(x, \tau)\dot{x}^5 \right]$$

$$= \frac{e^2}{4\pi c} \, e^{-|\tau - \tau_R|/\xi\lambda} \, \frac{\mathcal{F}^\mu_{\ \nu}(x, \tau)\dot{x}^\nu + c_5^2 \, \mathcal{F}^{5\mu}(x, \tau)}{1 + (c_5/c)^2},$$

where

$$\mathcal{F}^{\mu\nu}(x, \tau) = \frac{e}{4\pi R} \left\{ \frac{(z^\mu \beta^\nu - z^\nu \beta^\mu) \, \beta^2}{R^2} - \frac{\varepsilon \, (\tau - \tau_R)}{\xi\lambda c} \frac{z^\mu \beta^\nu - z^\nu \beta^\mu}{R} \right.$$

$$\left. - \frac{\left(z^\mu \dot{\beta}^\nu - z^\nu \dot{\beta}^\mu \right) R + (z^\mu \beta^\nu - z^\nu \beta^\mu) \left(\dot{\beta} \cdot z \right)}{R^2} \right\}$$

$$\mathcal{F}^{5\mu}(x, \tau) = \frac{e}{4\pi c R} \left\{ -\frac{z^\mu \beta^2 + \beta^\mu R}{R^2} - \frac{\varepsilon \, (\tau - \tau_R)}{\xi\lambda c} \frac{z^\mu + \beta^\mu Rc^2/c_5^2}{R} \right.$$

$$\left. + \frac{z^\mu \left(\dot{\beta} \cdot z \right)}{c R^2} \right\}.$$

In the limit $\lambda \to \infty$ and $c_5 \to 0$, we see that $c_5^2 \mathcal{F}^{5\mu}(x, \tau) \to 0$, and so the Lorentz force interaction reduces to the τ-independent expression

$$M\ddot{x}^\mu = \frac{e^2}{4\pi c} \, \mathcal{F}^\mu_{\ \nu}(x)\dot{x}^\nu$$

recovering the Lorentz force in the standard Maxwell form. The parameter c_5/c thus provides a continuous scaling of Maxwell's equations and the Lorentz force to the standard forms in Maxwell theory. The combined limit $\lambda \to \infty$ and $c_5 \to 0$ restricts the possible dynamics in SHP to those of Maxwell theory, as a system in τ-equilibrium [9].

4.9   BIBLIOGRAPHY

[1] Tanabashi, M., Hagiwara, K., Hikasa, K., Nakamura, K., Sumino, Y., Takahashi, F., Tanaka, J., Agashe, K., Aielli, G., Amsler, C., Antonelli, M., Asner, D. M., Baer, H., Banerjee, S., Barnett, R. M., Basaglia, T., Bauer, C. W., Beatty, J. J., Belousov, V. I., Beringer, J., Bethke, S., Bettini, A., Bichsel, H., Biebel, O., Black, K. M., Blucher, E., Buchmuller, O., Burkert, V., Bychkov, M. A., Cahn, R. N., Carena, M., Ceccucci, A., Cerri, A., Chakraborty, D., Chen, M. C., Chivukula, R. S., Cowan, G., Dahl, O., D'Ambrosio, G., Damour, T., de Florian, D., de Gouvêa, A., DeGrand, T., de Jong, P., Dissertori, G., Dobrescu, B. A., D'Onofrio, M., Doser, M., Drees, M., Dreiner, H. K., Dwyer, D. A., Eerola, P., Eidelman, S., Ellis, J., Erler, J., Ezhela, V. V., Fetscher, W., Fields, B. D., Firestone, R., Foster, B., Freitas, A., Gallagher, H., Garren, L., Gerber, H. J., Gerbier, G., Gershon, T., Gershtein, Y., Gherghetta, T., Godizov, A. A., Goodman, M., Grab, C., Gritsan, A. V., Grojean, C., Groom, D. E., Grünewald, M., Gurtu, A., Gutsche, T., Haber, H. E., Hanhart, C., Hashimoto, S., Hayato, Y., Hayes, K. G., Hebecker, A., Heinemeyer, S., Heltsley, B., Hernández-Rey, J. J., Hisano, J., Höcker, A., Holder, J., Holtkamp, A., Hyodo, T., Irwin, K. D., Johnson, K. F., Kado, M., Karliner, M., Katz, U. F., Klein, S. R., Klempt, E., Kowalewski, R. V., Krauss, F., Kreps, M., Krusche, B., Kuyanov, Y. V., Kwon, Y., Lahav, O., Laiho, J., Lesgourgues, J., Liddle, A., Ligeti, Z., Lin, C. J., Lippmann, C., Liss, T. M., Littenberg, L., Lugovsky, K. S., Lugovsky, S. B., Lusiani, A., Makida, Y., Maltoni, F., Mannel, T., Manohar, A. V., Marciano, W. J., Martin, A. D., Masoni, A., Matthews, J., Meißner, U. G., Milstead, D., Mitchell, R. E., Mönig, K., Molaro, P., Moortgat, F., Moskovic, M., Murayama, H., Narain, M., Nason, P., Navas, S., Neubert, M., Nevski, P., Nir, Y., Olive, K. A., Pagan, G. S., Parsons, J., Patrignani, C., Peacock, J. A., Pennington, M., Petcov, S. T., Petrov, V. A., Pianori, E., Piepke, A., Pomarol, A., Quadt, A., Rademacker, J., Raffelt, G., Ratcliff, B. N., Richardson, P., Ringwald, A., Roesler, S., Rolli, S., Romaniouk, A., Rosenberg, L. J., Rosner, J. L., Rybka, G., Ryutin, R. A., Sachrajda, C. T., Sakai, Y., Salam, G. P., Sarkar, S., Sauli, F., Schneider, O., Scholberg, K., Schwartz, A. J., Scott, D., Sharma, V., Sharpe, S. R., Shutt, T., Silari, M., Sjöstrand, T., Skands, P., Skwarnicki, T., Smith, J. G., Smoot, G. F., Spanier, S., Spieler, H., Spiering, C., Stahl, A., Stone, S. L., Sumiyoshi, T., Syphers, M. J., Terashi, K., Terning, J., Thoma, U., Thorne, R. S., Tiator, L., Titov, M., Tkachenko, N. P., Törnqvist, N. A., Tovey, D. R., Valencia, G., Van de Water, R., Varelas, N., Venanzoni, G., Verde, L., Vincter, M. G., Vogel, P., Vogt, A., Wakely, S. P., Walkowiak, W., Walter, C. W., Wands, D., Ward, D. R., Wascko, M. O., Weiglein, G., Weinberg, D. H., Weinberg, E. J., White, M., Wiencke, L. R., Willocq, S., Wohl, C. G., Womersley, J., Woody, C. L., Workman, R. L., Yao, W. M., Zeller, G. P., Zenin, O. V., Zhu, R. Y., Zhu, S. L., Zimmermann, F., Zyla, P. A., Anderson, J., Fuller, L., Lugovsky, V. S., and Schaffner, P. (Particle Data Group) 2018. *Physical Review D*, 98(3):030001. https://link.aps.org/doi/10.1103/PhysRevD.98.030001 49

[2] Land, M. 1996. *Foundations of Physics*, 27:19. 50

[3] Pierce, P. O. 1899. *A Short Table of Integrals*, Ginn and Company, New York. 52, 88

[4] Hestenes, D. 1966. *Space-Time Algebra*, Documents on modern physics, Gordon and Breach. https://books.google.co.il/books?id=OoRmatRYcs4C DOI: 10.1007/978-3-319-18413-5. 55

[5] Land, M. 2013. *Journal of Physics: Conference Series*, 437. https://doi.org/10.1088%2F1742-6596%2F437%2F1%2F012012 58

[6] Land, M. and Horwitz, L. 1991. *Foundations on Physics Letters*, 4:61. 62

[7] Jackson, J. 1975. *Classical Electrodynamics*, 9:391, Wiley, New York. DOI: 10.1063/1.3057859. 65

[8] Land, M. 2019. *Journal of Physics: Conference Series*, 1239:012005. https://doi.org/10.1088%2F1742-6596%2F1239%2F1%2F012005 65

[9] Land, M. 2017. *Journal of Physics: Conference Series*, 845:012024. http://stacks.iop.org/1742-6596/845/i=1/a=012024 69, 94

[10] Anderson, C. D. 1932. *Physical Review*, 41:405. 73

[11] Bethe, H. A. and Heitler, W. 1934. *Proc. Royal Society of London*, A(146):83. 73

[12] Horwitz, L. P. 2017 *Journal of Physics: Conference Series*, 845:012026. http://stacks.iop.org/1742-6596/845/i=1/a=012026 89

[13] Horwitz, L. P. and Arshansky, R. I. 2018. *Relativistic Many-Body Theory and Statistical Mechanics*, 2053–2571, Morgan & Claypool Publishers. http://dx.doi.org/10.1088/978-1-6817-4948-8 DOI: 10.1088/978-1-6817-4948-8. 89

CHAPTER 5

Advanced Topics

5.1 ELECTRODYNAMICS FROM COMMUTATION RELATIONS

In (2.1) we introduced an unconstrained 8D phase space (x^μ, p^μ) along with Poisson brackets for which

$$\{x^\mu, p^\nu\} = \frac{\partial x^\mu}{\partial x^\lambda}\frac{\partial p^\nu}{\partial p_\lambda} - \frac{\partial x^\mu}{\partial p_\lambda}\frac{\partial p^\nu}{\partial x^\lambda} = g^{\mu\nu}(x)$$

in curved spacetime. In 1990, Dyson [1] published a 1948 attempt by Feynman to derive the Lorentz force law and homogeneous Maxwell equations starting from Euclidean relations $\{x^i, p^j\} = \delta^{ij}$ on 6D phase space. Several authors noted that the derived equations have only Galilean symmetry, and so are not actually the Maxwell theory, leading to a number of interesting theoretical developments. Tanimura [2] generalized Feynman's derivation to Lorentz covariant form and obtained expressions similar to Maxwell theory, but including a fifth electromagnetic potential, a scalar evolution parameter that cannot be identified with proper time, absence of reparameterization invariance, and violations of the mass-shell constraint. His result can be identified with SHP electrodynamics. Significantly, Hojman and Shepley [3] proved that the existence of quantum commutation relations is a strong assumption, sufficient to determine a corresponding classical action, from which this system can be derived. We generalize Tanimura's result to curved spacetime and show that this approach to SHP provides the final step in Feynman's program. Using the technique of Hojman and Shepley, we show that SHP electrodynamics follows as the most general interacting system consistent with the unconstrained commutation relations we have assumed [4].

We begin with the commutation relations among the quantum operators

$$[x^\mu, x^\nu] = 0 \qquad m[x^\mu, \dot{x}^\nu] = i\hbar g^{\mu\nu}(x) \qquad (5.1)$$

for $\mu, \nu = 0, 1, \cdots, D-1$, and suppose equations of motion

$$m\ddot{x}^\mu = F^\mu(\tau, x, \dot{x}).$$

We regard these quantities as operators in a Heisenberg picture, so that the field equations and the Lorentz force may be interpreted, in the Ehrenfest sense, as relations among the expectation values which correspond to relations among classical quantities. It follows that

$$[\dot{x}^\mu, q(x)] = \frac{i\hbar}{m}\frac{\partial q}{\partial x_\mu} \qquad (5.2)$$

for any function $q(x)$. Differentiating (5.1) with respect to τ we find

$$m[\dot{x}^\mu, \dot{x}^\nu] + m[x^\mu, \ddot{x}^\nu] = i\hbar\partial_\rho g^{\mu\nu}(x)\dot{x}^\rho$$

and so define $W^{\mu\nu} = -W^{\nu\mu}$ by

$$W^{\mu\nu} = \frac{m^2}{i\hbar}[\dot{x}^\mu, \dot{x}^\nu]. \tag{5.3}$$

From (5.1) and the Jacobi identity,

$$[x^\lambda, [\dot{x}^\mu, \dot{x}^\nu]] + [\dot{x}^\mu, [\dot{x}^\nu, x^\lambda]] + [\dot{x}^\nu, [x^\lambda, \dot{x}^\nu]] = 0$$

we find that

$$[x^\lambda, W^{\mu\nu}] = \frac{m^2}{i\hbar}\left([[x^\lambda, \dot{x}^\mu], \dot{x}^\nu] + [\dot{x}^\mu, [x^\lambda, \dot{x}^\nu]]\right) = i\hbar\left(\partial^\nu g^{\lambda\mu} - \partial^\mu g^{\lambda\nu}\right).$$

Defining $f^{\mu\nu} = -f^{\nu\mu}$ by

$$f^{\mu\nu} = W^{\mu\nu} - m\left\langle(\partial^\nu g^{\lambda\mu} - \partial^\mu g^{\lambda\nu})\dot{x}_\lambda\right\rangle, \tag{5.4}$$

where the brackets $\langle...\rangle$ represent Weyl ordering, we find

$$[x^\sigma, f^{\mu\nu}] = 0,$$

which shows that $f^{\mu\nu}$ is independent of $\dot{x}$. When lowering indices, we define

$$\dot{x}_\mu = \langle g_{\mu\nu}(x)\dot{x}^\nu\rangle$$

and from

$$[\dot{x}_\mu, \dot{x}_\nu] = \left[\left\langle g_{\mu\lambda}\dot{x}^\lambda\right\rangle, \left\langle g_{\nu\rho}\dot{x}^\rho\right\rangle\right]$$

we may show that

$$f_{\mu\nu} = g_{\mu\lambda}g_{\nu\rho}f^{\lambda\rho} = -\frac{m^2}{i\hbar}[\dot{x}_\mu, \dot{x}_\nu] \tag{5.5}$$

leading to the Bianchi relation

$$\partial_\mu f_{\nu\rho} + \partial_\nu f_{\rho\mu} + \partial_\rho f_{\mu\nu} = 0.$$

Rearranging Equation (5.1) and using (5.3) and (5.4), we see that

$$m[x^\mu, \ddot{x}^\nu] = \frac{i\hbar}{m}f^{\mu\nu} + 2i\hbar\langle\Gamma^{\nu\lambda\mu}\dot{x}_\lambda\rangle,$$

where

$$\Gamma^{\nu\lambda\mu} = -\frac{1}{2}(\partial^\mu g^{\lambda\nu} + \partial^\lambda g^{\mu\nu} - \partial^\nu g^{\lambda\mu})$$

is the Levi–Civita connection.

We now define g^μ through the equation

$$F^\mu = m\ddot{x}^\mu = g^\mu(x, \dot{x}, \tau) + \langle f^{\mu\nu}\dot{x}_\nu \rangle - m\langle \Gamma^{\mu\lambda\nu}\dot{x}_\lambda\dot{x}_\nu \rangle$$

and it follows that

$$
\begin{aligned}
[x^\lambda, g^\mu] &= [x^\lambda, f^\mu] - f^{\mu\nu}[x^\lambda, \dot{x}_\nu] + m\,\Gamma^{\mu\nu\rho}[x^\lambda, \dot{x}_\nu]\,\dot{x}_\rho + \Gamma^{\mu\nu\rho}\,\dot{x}_\nu[x^\lambda, \dot{x}_\rho] \\
&= \frac{i\hbar}{m}\,f^{\lambda\mu} + 2i\hbar\left\langle \Gamma^{\mu\rho\lambda}\dot{x}_\rho \right\rangle + \frac{i\hbar}{m}\,f^{\mu\nu}\,\delta^\lambda_\nu - i\hbar\left\langle \left(\Gamma^{\mu\nu\rho}\,\delta^\lambda_\nu\,\dot{x}_\rho + \Gamma^{\mu\nu\rho}\,\dot{x}_\nu\,\delta^\lambda_\rho\right)\right\rangle \\
&= 0
\end{aligned}
$$

so that g^μ is also independent of $\dot{x}$. We may write the force as

$$g^\mu + \langle f^{\mu\nu}\dot{x}_\nu \rangle = m\left[\ddot{x}^\mu + \left\langle \Gamma^{\mu\lambda\nu}\dot{x}_\lambda\dot{x}_\nu \right\rangle\right] = m\frac{D\dot{x}^\mu}{D\tau}$$

and since

$$m\ddot{x}^\mu = m\frac{d}{d\tau}\langle g^{\mu\nu}\dot{x}_\nu \rangle\,,$$

we lower the index of g^μ to find

$$g_\nu = g_{\nu\lambda}\,f^\lambda - \langle g_{\nu\lambda}\,f^{\lambda\rho}\,\dot{x}_\rho \rangle + m\langle g_{\nu\lambda}\,\Gamma^{\lambda\rho\sigma}\,\dot{x}_\rho\dot{x}_\sigma \rangle.$$

We write the first term on the right-hand side as

$$g_{\nu\lambda}\,f^\lambda = m\,\langle g_{\nu\lambda}\,\ddot{x}^\lambda \rangle = m\,g_{\nu\lambda}\frac{d}{d\tau}\langle g^{\lambda\rho}\dot{x}_\rho \rangle = m\,\ddot{x}_\nu + m\,\langle g_{\nu\lambda}\,\partial^\sigma g^{\lambda\rho}\,\dot{x}_\rho\dot{x}_\sigma \rangle.$$

Since the indices ρ and σ of $\partial^\sigma g^{\lambda\rho}$ occur in symmetric combination, we may write

$$\frac{1}{2}\left(\partial^\sigma g^{\lambda\rho} + \partial^\rho g^{\lambda\sigma}\right) = -\Gamma^{\lambda\rho\sigma} + \frac{1}{2}\partial^\lambda g^{\rho\sigma}$$

so that

$$g_\nu = m\,\ddot{x}_\nu + \frac{1}{2}m\,\langle \partial_\nu g^{\lambda\rho}\,\dot{x}_\lambda\dot{x}_\rho \rangle - \langle f_{\nu\lambda}\,g^{\lambda\rho}\,\dot{x}_\rho \rangle.$$

Using (5.2) and (5.5) we obtain

$$
\begin{aligned}
[\dot{x}_\mu, g_\nu] &= m[\dot{x}_\mu, \ddot{x}_\nu] + \left\langle \frac{1}{2}i\hbar\partial_\mu\partial_\nu g^{\lambda\rho}\,\dot{x}_\lambda\dot{x}_\rho - \frac{i\hbar}{2m}\partial_\nu g^{\lambda\rho}(f_{\mu\lambda}\dot{x}_\rho + \dot{x}_\lambda f_{\mu\rho}) \right. \\
&\qquad \left. - \frac{i\hbar}{m}\partial_\mu(f_{\nu\lambda}\,g^{\lambda\rho})\,\dot{x}_\rho + \frac{i\hbar}{m^2}\,f_{\nu\lambda}\,g^{\lambda\rho}\,f_{\mu\rho} \right\rangle \\
&= m[\dot{x}_\mu, \ddot{x}_\nu] + \left\langle \frac{1}{2}i\hbar\partial_\mu\partial_\nu g^{\lambda\rho}\,\dot{x}_\lambda\dot{x}_\rho - \frac{i\hbar}{m}(\partial_\nu g^{\lambda\rho})f_{\mu\lambda}\,\dot{x}_\rho \right. \\
&\qquad \left. - \frac{i\hbar}{m}(\partial_\mu g^{\lambda\rho})f_{\nu\lambda}\,\dot{x}_\rho - \frac{i\hbar}{m}\partial_\mu f_{\nu\lambda}g^{\lambda\rho}\,\dot{x}_\rho + \frac{i\hbar}{m^2}\,f_{\nu\lambda}\,g^{\lambda\rho}\,f_{\mu\rho} \right\rangle.
\end{aligned}
$$

Finally, antisymmetrization with respect to the indices μ and ν gives

$$\begin{aligned}
[\dot{x}_\mu, g_\nu] - [\dot{x}_\nu, g_\mu] &= m[\dot{x}_\mu, \ddot{x}_\nu] - [\dot{x}_\nu, \ddot{x}_\mu] - \frac{i\hbar}{m}\left\langle (\partial_\mu f_{\nu\lambda} - \partial_\nu f_{\mu\lambda}) g^{\lambda\rho} \dot{x}_\rho \right\rangle \\
&= m\frac{d}{d\tau}[\dot{x}_\mu, \dot{x}_\nu] - \frac{i\hbar}{m}\left\langle (\partial_\mu f_{\nu\lambda} + \partial_\nu f_{\lambda\mu}) \dot{x}^\lambda \right\rangle \\
&= -\frac{i\hbar}{m}\frac{d}{d\tau} f_{\mu\nu} - \frac{i\hbar}{m}\left\langle (\partial_\mu f_{\nu\rho} + \partial_\nu f_{\rho\mu}) \dot{x}^\rho \right\rangle \\
&= -\frac{i\hbar}{m}\left[\langle (\partial_\rho f_{\mu\nu} + \partial_\mu f_{\nu\rho} + \partial_\nu f_{\rho\mu}) \dot{x}^\rho \rangle + \partial_\tau f_{\mu\nu} \right]
\end{aligned}$$

and so using the Bianchi identity for $f_{\mu\nu}$,

$$\partial_\mu g_\nu - \partial_\nu g_\mu + \frac{\partial f_{\mu\nu}}{\partial\tau} = 0.$$

Regarding these equations in the Ehrenfest sense, we may summarize the classical theory as

$$m\frac{D\dot{x}^\mu}{D\tau} = m[\ddot{x}^\mu + \sigma^{\mu\lambda\nu} \dot{x}_\lambda \dot{x}_\nu] = f^{\mu\nu} \dot{x}_\nu + g^\mu$$

$$\partial_\mu f_{\nu\rho} + \partial_\nu f_{\rho\mu} + \partial_\rho f_{\mu\nu} = 0$$

$$\partial_\mu g_\nu - \partial_\nu g_\mu + \frac{\partial f_{\mu\nu}}{\partial\tau} = 0.$$

Introducing the definitions

$$x^D = \tau \qquad \partial_\tau = \partial_D \qquad f_{\mu D} = -f_{D\mu} = g_\mu \,.$$

We may then combine the inhomogeneous field equations as

$$\partial_\alpha f_{\beta\gamma} + \partial_\beta f_{\gamma\alpha} + \partial_\gamma f_{\alpha\beta} = 0 \tag{5.6}$$

(for $\alpha, \beta, \gamma = 0, \cdots, D$), which shows that the two form f is closed on the formal $(D+1)$-dimensional manifold (τ, x). Hence, if this manifold is contractable, then f is an exact form which can be obtained as the derivative of some potential with the form $f = da$. The Lorentz force equation becomes

$$m\frac{D\dot{x}^\mu}{D\tau} = m[\ddot{x}^\mu + \Gamma^{\mu\lambda\nu} \dot{x}_\lambda \dot{x}_\nu] = f^{\mu\nu}(\tau, x)\dot{x}_\nu + g^\mu(\tau, x) = f^\mu{}_\beta(\tau, x)\dot{x}^\beta \,. \tag{5.7}$$

Following Dyson and Feynman, we observe that given Equation (5.6), the two-form $f^{\alpha\beta}$ is determined if we know functions ρ and j^μ such that

$$\mathcal{D}_\alpha f^{\mu\alpha} = j^\mu \qquad\qquad \mathcal{D}_\alpha f^{d\alpha} = \rho,$$

where $\mathcal{D}_\alpha$ is the covariant derivative. By denoting $\rho = j^d$, these equations can be written compactly as

$$\mathcal{D}_\alpha \, f^{\beta\alpha} = j^\beta, \tag{5.8}$$

where, due to the antisymmetry of $f^{\beta\alpha}$, we see that j^β is conserved as $\mathcal{D}_\alpha j^\alpha = 0$.

$$\mathcal{D}_\alpha j^\alpha = 0 \,.$$

By comparing the Lorentz force (5.7) with (3.6), and the field Equations (5.6) and (5.8) with (3.19) and (3.17), we see conclude that the assumption of unconstrained commutation relations leads to a field theory equivalent to classical SHP electrodynamics.

In Sections 3.2 and 3.3 we found the Lorentz force and field equations from an action principle. Hojman and Shepley [3] set out to prove that the assumed commutation relations are sufficient to establish the existence of a unique Lagrangian of electromagnetic form. To accomplish this goal, they demonstrate a new connection between the commutation relations and well-established results from the inverse problem in the calculus of variations, a theory which concerns the conditions under which a system of differential equations may be derived from a variational principle. We consider a set of ordinary second-order differential equations of the form

$$F_k(\tau, q, \dot{q}, \ddot{q}) = 0 \qquad \dot{q}^j = \frac{dq^j}{d\tau} \qquad \ddot{q}^j = \frac{d^2 q^j}{d\tau^2} \qquad j, k = 1, \cdots, n.$$

Under variations of the path

$$\begin{aligned}
q(\tau) &\longrightarrow q(\tau) + dq(\tau) \\
\dot{q}(\tau) &\longrightarrow \dot{q}(\tau) + d\dot{q}(\tau) = \dot{q}(\tau) + \frac{d}{d\tau} dq(\tau) \\
\ddot{q}(\tau) &\longrightarrow \ddot{q}(\tau) + d\ddot{q}(\tau) = \ddot{q}(\tau) + \frac{d^2}{d\tau^2} dq(\tau)
\end{aligned}$$

the function $F_k(\tau, q, \dot{q}, \ddot{q})$ admits the variational one-form defined by

$$dF_k = \frac{\partial F_k}{\partial q^j} dq^j + \frac{\partial F_k}{\partial \dot{q}^j} d\dot{q}^j + \frac{\partial F_k}{\partial \ddot{q}^j} d\ddot{q}^j$$

and the variational two-form

$$dq^k dF_k = \frac{\partial F_k}{\partial q^j} dq^k \wedge dq^j + \frac{\partial F_k}{\partial \dot{q}^j} dq^k \wedge d\dot{q}^j + \frac{\partial F_k}{\partial \ddot{q}^j} dq^k \wedge d\ddot{q}^j,$$

where the $3n$ path variations $(dq^k, d\dot{q}^k, d\ddot{q}^k)$ for $k = 1, \cdots, n$ are understood to be linearly independent. The system of differential equations $F_k(\tau, q, \dot{q}, \ddot{q})$ is called self-adjoint if there exists a two-form $\Omega_2(dq, d\dot{q})$ such that for all admissible variations of the path,

$$dq^k dF_k(dq) = \frac{d}{d\tau} \Omega_2(dq, d\dot{q}).$$

Through integration by parts, one may show [5] that such a two-form exists and is unique up to an additive constant, if and only if

$$\frac{\partial F_i}{\partial \ddot{q}^k} = \frac{\partial F_k}{\partial \ddot{q}^i} \qquad (5.9)$$

$$\frac{\partial F_i}{\partial \dot{q}^k} + \frac{\partial F_k}{\partial \dot{q}^i} = \frac{d}{d\tau}\left[\frac{\partial F_i}{\partial \ddot{q}^k} + \frac{\partial F_k}{\partial \ddot{q}^i}\right] \qquad (5.10)$$

$$\frac{\partial F_i}{\partial q^k} - \frac{\partial F_k}{\partial q^i} = \frac{1}{2}\frac{d}{d\tau}\left[\frac{\partial F_i}{\partial \dot{q}^k} - \frac{\partial F_k}{\partial \dot{q}^i}\right], \qquad (5.11)$$

known as the Helmholtz conditions [6, 7]. Introducing the notation

$$\delta = dq_\beta^k \frac{\partial}{\partial q_\beta^k} \qquad q_\beta^k = \left(\frac{d}{d\tau}\right)^\beta q^k \qquad \beta = 0, 1, 2,$$

it follows that

$$\delta^2 = dq_\beta^k \wedge dq_\alpha^l \frac{\partial^2}{\partial q_\beta^k \partial q_\alpha^l} = 0,$$

which permits the equivalence of a set of self-adjointness differential equations to a Lagrangian formulation to be easily demonstrated [8]. Varying the Lagrangian L,

$$\delta L = \frac{\partial L}{\partial q^k}dq^k + \frac{\partial L}{\partial \dot{q}^k}d\dot{q}^k = \left[-\frac{d}{d\tau}\frac{\partial L}{\partial \dot{q}^k} + \frac{\partial L}{\partial q^k}\right]dq^k + \frac{d}{d\tau}\left(\frac{\partial L}{\partial \dot{q}^k}dq^k\right) = F_k dq^k + \frac{d}{d\tau}\Omega_1$$

so that

$$\delta^2 = 0 \implies -dq^k \delta F_k + \frac{d}{d\tau}\delta\Omega_1 = -dq^k \delta F_k + \frac{d}{d\tau}\Omega_2 = 0$$

which demonstrates self-adjointness. Conversely, self-adjoint of F_k requires that $dq^k \delta F_k - \frac{d}{d\tau}\Omega_2 = 0$ and since $\delta^2 = 0$,

$$\frac{d}{d\tau}\Omega_2 = \delta\frac{d}{d\tau}\Omega_1.$$

Therefore,

$$0 = dq^k \delta F_k - \frac{d}{d\tau}\Omega_2 = \delta\left(dq^k F_k - \frac{d}{d\tau}\Omega_1\right) = \delta L$$

by variation of L under τ-integration, one obtains the differential equations $F_k = 0$.

For the second-order equations considered here, it follows [5] from self-adjointness that the most general form of F_k is

$$F_k(\tau, q, \dot{q}, \ddot{q}) = A_{kj}(\tau, q, \dot{q})\ddot{q}^j + B_k(\tau, q, \dot{q}). \qquad (5.12)$$

To see this, notice that F_k is independent of $d^3 q^i / dt^3$, so that the right-hand side of (5.10) must be independent of $\dddot{q}^i$. Inserting (5.12) into (5.9)–(5.11), one finds the Helmholtz conditions on A_{kj} and B_k

$$A_{ij} = A_{ji} \qquad\qquad \frac{\partial A_{ij}}{\partial \dot{q}^k} = \frac{\partial A_{kj}}{\partial \dot{q}^i}$$

$$\frac{\partial B_i}{\partial \dot{q}^j} + \frac{\partial B_j}{\partial \dot{q}^i} = 2\left[\frac{\partial}{\partial \tau} + \dot{q}^k \frac{\partial}{\partial q^k}\right] A_{ij}$$

$$\frac{\partial B_i}{\partial q^j} - \frac{\partial B_j}{\partial q^i} = \frac{1}{2}\left[\frac{\partial}{\partial \tau} + \dot{q}^k \frac{\partial}{\partial q^k}\right]\left(\frac{\partial B_i}{\partial \dot{q}^j} - \frac{\partial B_j}{\partial \dot{q}^i}\right)$$

along with the useful identity

$$\frac{\partial A_{ij}}{\partial q^k} - \frac{\partial A_{kj}}{\partial q^i} = \frac{1}{2}\frac{\partial}{\partial \dot{q}^j}\left(\frac{\partial B_i}{\partial \dot{q}^k} - \frac{\partial B_k}{\partial \dot{q}^i}\right) . \tag{5.13}$$

In the domain of invertibilty of the A_{jk}, one can write (5.12) as

$$F_k(\tau, q, \dot{q}, \ddot{q}) = A_{kj}(\tau, q, \dot{q})[\ddot{q}^j - f^j] \qquad\qquad f^j(\tau, q, \dot{q}) = -(A^{-1})^{jk} B_k$$

and the Helmholtz conditions become

$$A_{ij} = A_{ji} \qquad\qquad \frac{\partial A_{ij}}{\partial \dot{q}^k} = \frac{\partial A_{kj}}{\partial \dot{q}^i} \tag{5.14}$$

$$\frac{D}{D\tau} A_{ij} = -\frac{1}{2}\left[A_{ik}\frac{\partial f^k}{\partial \dot{q}^j} + A_{jk}\frac{\partial f^k}{\partial \dot{q}^i}\right] \tag{5.15}$$

$$\frac{1}{2}\frac{D}{D\tau}\left[A_{ik}\frac{\partial f^k}{\partial \dot{q}^j} - A_{jk}\frac{\partial f^k}{\partial \dot{q}^i}\right] = A_{ik}\frac{\partial f^k}{\partial q^j} - A_{jk}\frac{\partial f^k}{\partial q^i},$$

where

$$\frac{D}{D\tau} = \frac{\partial}{\partial \tau} + \dot{q}^k \frac{\partial}{\partial q^k} + f^k \frac{\partial}{\partial \dot{q}^k}$$

is the total time derivative subject to the constraint

$$\ddot{q}^k - f^k(\tau, q, \dot{q}) = 0 . \tag{5.16}$$

The identity (5.13) becomes

$$\frac{\partial A_{ij}}{\partial q^k} - \frac{\partial A_{kj}}{\partial q^i} = -\frac{1}{2}\frac{\partial}{\partial \dot{q}^j}\left[\frac{\partial}{\partial \dot{q}^k}(A_{in}f^n) - \frac{\partial}{\partial \dot{q}^i}(A_{kn}f^n)\right]. \tag{5.17}$$

Within the domain of applicability of the inverse function theorem, (5.16) is equivalent to (5.12), and the Helmholtz conditions become the necessary and sufficient conditions for the existence of an integrating factor A_{jk} such that

$$F_k = A_{kj}(\tau, q, \dot{q})[\ddot{q}^j - f^j] = \frac{d}{d\tau}\left(\frac{\partial L}{\partial \dot{q}^k}\right) - \frac{\partial L}{\partial q^k}. \tag{5.18}$$

Employing this apparatus, Hojman and Shepley prove that given the quantum mechanical commutation relations

$$[x^i(\tau), \dot{x}^j(\tau)] = i\hbar G^{ij},$$

the classical function

$$g^{ij} = \lim_{\hbar \to 0} G^{ij}$$

has an inverse

$$\omega_{ij} = (g^{-1})_{ij}$$

which satisfies the Helmholtz conditions. Following Santilli, we take the function $A_{\mu\nu} = g_{\mu\nu}(x)$ to be a Riemannian metric independent of $\dot{x}$, so that Equation (5.14) is satisfied automatically. Since $g_{\mu\nu}$ does not depend on $\dot{x}^\mu$, Equation (5.15) becomes

$$\frac{D}{D\tau}g_{\mu\nu} = \dot{x}^\sigma\frac{\partial}{\partial x^\sigma}g_{\mu\nu} = -\frac{1}{2}\left[\frac{\partial f_\mu}{\partial \dot{x}^\nu} - \frac{\partial f_\nu}{\partial \dot{x}^\mu}\right] \tag{5.19}$$

and Equation (5.17) becomes

$$-\frac{1}{2}\frac{\partial}{\partial \dot{x}^\nu}\left[\frac{\partial f_\mu}{\partial \dot{x}^\sigma} - \frac{\partial f_\sigma}{\partial \dot{x}^\mu}\right] = \frac{\partial g_{\mu\nu}}{\partial x^\sigma} - \frac{\partial g_{\sigma\nu}}{\partial x^\mu}. \tag{5.20}$$

Acting on (5.19) with $\partial/\partial\dot{x}^\sigma$ and exchanging $(\nu \leftrightarrow \sigma)$, we obtain

$$g_{\mu\sigma,\nu} = -\frac{1}{2}\left[\frac{\partial^2 f_\mu}{\partial\dot{x}^\sigma\partial\dot{x}^\nu} + \frac{\partial^2 f_\sigma}{\partial\dot{x}^\mu\partial\dot{x}^\nu}\right], \tag{5.21}$$

where $g_{\mu\sigma,\nu} = \partial g_{\mu\sigma}/\partial x^\nu$. Combining (5.20) and (5.21), we find

$$\frac{1}{2}\frac{\partial^2 f_\mu}{\partial\dot{x}^\sigma\partial\dot{x}^\nu} = -\frac{1}{2}(g_{\mu\nu,\sigma} + g_{\mu\sigma,\nu} - g_{\sigma\nu,\mu}) = -\Gamma_{\mu\sigma\nu},$$

where $\Gamma_{\mu\sigma\nu}$ is the Levi-Civita connection. Thus, the most general expression for $f_\mu(\tau, x, \dot{x})$ is

$$f_\mu = -\Gamma_{\mu\nu\sigma}\dot{x}^\nu\dot{x}^\sigma - \rho_{\mu\nu}(\tau, x)\dot{x}^\nu - \sigma_\mu(\tau, x). \tag{5.22}$$

Now from (5.19) we find

$$\dot{x}^\sigma\frac{\partial g_{\mu\nu}}{\partial x^\sigma} = \frac{1}{2}\left[2\Gamma_{\mu\nu\sigma}\dot{x}^\sigma + 2\Gamma_{\nu\mu\sigma}\dot{x}^\sigma + \rho_{\mu\nu} + \rho_{\nu\mu}\right]$$

and using

$$(\Gamma_{\mu\nu\sigma} + \Gamma_{\nu\mu\sigma})\dot{x}^\sigma = g_{\mu\nu,\sigma}\dot{x}^\sigma$$

we find that all terms except for those in $\rho_{\mu\nu}$ cancel, so that

$$0 = \rho_{\mu\nu} + \rho_{\nu\mu} .$$

We now apply Equation (5.16) which becomes

$$\frac{1}{2}\frac{D}{D\tau}\left[g_{\mu\sigma}\frac{\partial f^\sigma}{\partial \dot{x}^\nu} - g_{\nu\sigma}\frac{\partial f^\sigma}{\partial \dot{x}^\mu}\right] = g_{\mu\sigma}\frac{\partial f^\sigma}{\partial x^\nu} - g_{\nu\sigma}\frac{\partial f^\sigma}{\partial x^\mu}$$

$$\frac{1}{2}\frac{D}{D\tau}\left[\frac{\partial f_\mu}{\partial \dot{x}^\nu} - \frac{\partial f_\nu}{\partial \dot{x}^\mu}\right] = f_{\mu,\nu} - f_{\nu,\mu} - g_{\mu\sigma,\nu}f^\sigma + g_{\nu\sigma,\mu}f^\sigma. \qquad (5.23)$$

using (5.22) to expand the left-hand side,

$$\frac{1}{2}\frac{D}{D\tau}\left[\frac{\partial f_\mu}{\partial \dot{x}^\nu} - \frac{\partial f_\nu}{\partial \dot{x}^\mu}\right] = -\frac{1}{2}\frac{D}{D\tau}\left[\frac{\partial}{\partial \dot{x}^\nu}\Big(\Gamma_{\mu\lambda\sigma}\dot{x}^\lambda\dot{x}^\sigma + \rho_{\mu\lambda}(\tau,x)\dot{x}^\lambda \right.$$

$$\left. + \sigma_\mu(\tau,x)\Big) - (\mu \leftrightarrow \nu)\right]$$

$$= -\frac{1}{2}\frac{D}{D\tau}\left[2(\Gamma_{\mu\nu\lambda} - \Gamma_{\nu\mu\lambda})\dot{x}^\lambda + \rho_{\mu\nu} - \rho_{\nu\mu}\right]$$

$$= -\left(\frac{\partial}{\partial \tau} + \dot{x}^\sigma\frac{\partial}{\partial x^\sigma} + f^\sigma\frac{\partial}{\partial \dot{x}^\sigma}\right)\left[(g_{\mu\lambda,\nu} - g_{\nu\lambda,\mu})\dot{x}^\lambda + \rho_{\mu\nu}\right]$$

$$= -(g_{\mu\sigma,\nu} - g_{\nu\sigma,\mu})f^\sigma - \rho_{\mu\nu,\tau}$$

$$- \dot{x}^\lambda\dot{x}^\sigma(g_{\mu\lambda,\nu\sigma} - g_{\nu\lambda,\mu\sigma}) + \dot{x}^\lambda\rho_{\mu\nu,\lambda}, \qquad (5.24)$$

where $\rho_{\mu\nu,\tau} = \partial\rho_{\mu\nu}/\partial\tau$, and we have used

$$2(\Gamma_{\mu\nu\lambda} - \Gamma_{\nu\mu\lambda})\dot{x}^\lambda = \dot{x}^\lambda(-g_{\nu\lambda,\mu} + g_{\mu\lambda,\nu} + g_{\nu\mu,\lambda} + g_{\mu\lambda,\nu} - g_{\nu\lambda,\mu} - g_{\mu\nu,\lambda})$$

$$= 2\dot{x}^\lambda(g_{\mu\lambda,\nu} - g_{\nu\lambda,\mu}) .$$

Again using (5.22) we have

$$f_{\mu,\nu} = -\left[\Gamma_{\mu\lambda\sigma}\dot{x}^\lambda\dot{x}^\sigma + \rho_{\mu\lambda}(\tau,x)\dot{x}^\lambda + \sigma_\mu(\tau,x)\right]_{,\nu}$$

$$= -\left[\Gamma_{\mu\lambda\sigma,\nu}\dot{x}^\lambda\dot{x}^\sigma + \rho_{\mu\lambda,\nu}\dot{x}^\lambda + \sigma_{\mu,\nu}\right]$$

so that the right-hand side of (5.23) is

$$f_{\mu,\nu} - f_{\nu,\mu} - g_{\mu\sigma,\nu}f^\sigma + g_{\nu\sigma,\mu}f^\sigma =$$

$$= -\left[(\Gamma_{\mu\lambda\sigma,\nu} - \Gamma_{\nu\lambda\sigma,\mu})\dot{x}^\lambda\dot{x}^\sigma + (\rho_{\mu\lambda,\nu} - \rho_{\nu\lambda,\mu})\dot{x}^\lambda + \sigma_{\mu,\nu} - \sigma_{\nu,\mu}\right]$$

$$- (g_{\mu\sigma,\nu} - g_{\nu\sigma,\mu})f^\sigma.$$

Now canceling common terms, we are left with

$$\frac{\partial \rho_{\mu\nu}}{\partial \tau} + \dot{x}^\lambda \rho_{\mu\nu,\lambda} = \dot{x}^\lambda (\rho_{\mu\lambda,\nu} - \rho_{\nu\lambda,\mu}) + \sigma_{\mu,\nu} - \sigma_{\nu,\mu}$$

which, because the $\dot{x}^\lambda$ are arbitrary, is equivalent the two expressions

$$\frac{\partial \rho_{\mu\nu}}{\partial \tau} = \frac{\partial \sigma_\mu}{\partial x^\nu} - \frac{\partial \sigma_\nu}{\partial x^\mu} \qquad \partial_\lambda \rho_{\mu\nu} + \partial_\mu \rho_{\nu\lambda} + \partial_\nu \rho_{\lambda\mu} = 0.$$

Therefore, we may identify

$$f_{\mu\nu} = -\rho_{\mu\nu} \qquad \text{and} \qquad f_{5\mu} = -\sigma_\mu$$

showing that SHP electrodynamics is the most general interaction consistent with the unconstrained commutation relations.

Moreover, these commutation relations are sufficient to establish the existence of an equivalent Lagrangian for the classical problem associated with the quantum commutators. We observe that in flat space (5.18) implies

$$\eta_{\mu\nu}[M \, \ddot{x}^\nu - f^\nu] = \frac{d}{d\tau}\left(\frac{\partial L}{\partial \dot{x}^\mu}\right) - \frac{\partial L}{\partial x^\mu}$$

$$= \frac{\partial^2 L}{\partial \dot{x}^\mu \partial \dot{x}^\nu} \ddot{x}^\nu + \frac{\partial^2 L}{\partial \dot{x}^\mu \partial x^\nu} \dot{x}^\nu + \frac{\partial^2 L}{\partial \dot{x}^\mu \partial \tau} - \frac{\partial L}{\partial x^\mu}$$

so that the solution

$$M \, \eta_{\mu\nu} = \frac{\partial^2 L}{\partial \dot{x}^\mu \partial \dot{x}^\nu} \qquad -\eta_{\mu\nu} f^\nu = \frac{\partial^2 L}{\partial \dot{x}^\mu \partial x^\nu} \dot{x}^\nu + \frac{\partial^2 L}{\partial \dot{x}^\mu \partial \tau} - \frac{\partial L}{\partial x^\mu}$$

is unique. Therefore, we see that L may consist of the quadratic term integrated from the first expression, plus terms at most linear in $\dot{x}^\mu$. Thus, we may write the Lagrangian

$$L = \frac{1}{2} M \, \dot{x}^\mu \dot{x}_\mu + \frac{e}{c} a_\mu(\tau, x) \dot{x}^\mu + \frac{e c_5}{c} a_5$$

which is the SHP event Lagrangian (3.3) in flat space. This demonstrates that SHP electrodynamics represents the conditions on the most general velocity dependent forces that may be obtained from a variational principle.

5.2 CLASSICAL NON-ABELIAN GAUGE THEORY

A classical non-Abelian gauge theory was given by Wong [9] possessing the following structure:

$$m\ddot{\xi}_\mu = g\mathbf{F}_{\mu\nu} \cdot \mathbf{I}(\tau) \, \dot{\xi}^\nu$$
$$\dot{\mathbf{I}} = -g\mathbf{A}_\mu \times \mathbf{I} \, \dot{\xi}^\mu$$
$$\mathbf{F}_{\mu\nu} = \partial_\mu \mathbf{A}_\nu - \partial_\nu \mathbf{A}_\mu + g\mathbf{A}_\mu \times \mathbf{A}_\nu$$
$$\partial^\mu \mathbf{F}_{\mu\nu} + g\mathbf{A}^\mu \times \mathbf{F}_{\mu\nu} = -\mathbf{j}_\nu$$
$$\mathbf{A}^\mu = A_{a\mu} I^a \qquad \mathbf{F}_{\mu\nu} = F_{a\mu\nu} I^a \qquad [I^a, I^b] = i\hbar \varepsilon^{abc} I_c,$$

where $\xi^\mu(\tau)$ is the particle worldline and the $I^a(\tau)$ are an operator representation of the generators of a non-Abelian gauge group. From the form of the field $\mathbf{f}_{\mu\nu}$, one has the inhomogeneous equation

$$\mathcal{D}_\mu \mathbf{F}_{\nu\rho} + \mathcal{D}_\nu \mathbf{F}_{\rho\mu} + \mathcal{D}_\rho \mathbf{F}_{\mu\nu} = 0$$

with covariant derivative

$$(\mathcal{D}_\mu \mathbf{F}_{\mu\nu})_a = \partial_\mu F_{a\mu\nu} - \varepsilon_a^{\ bc} A_{b\mu} F_{c\mu\nu} \ .$$

Lee [10] followed Feynman's method, supplementing the phase space commutation relations with

$$[I^a, I^b] = i\hbar\varepsilon^{abc} I_c \qquad [x_i, I^a(t)] = 0 \qquad \dot{\mathbf{I}} + g\mathbf{A}_i \times \mathbf{I}\,\dot{x}^i = 0$$

for $i = 1, 2, 3$, and arrived at the Wong's equations in Newtonian form. Tanimura [2] generalized Lee's derivation to D-dimensional flat Minkowski space and a general gauge group satisfying

$$[I^a, I^b] = i\hbar\,f_c^{\ ab} I^c \qquad \dot{I}^a = F_c^{\ ab} A_{b\mu}(x)\,\dot{x}^\mu I^c \qquad (5.25)$$

for τ-independent fields.

We now extend the presentation of Section 5.1 by generalizing the Helmholtz conditions to take account of classical non-Abelian gauge fields according to Wong's formulation. To achieve this, we associate with variations dq of the path $q(\tau)$, a variation dI^a of the generators I^a, which may be understood as the variation of the orientation of the tangent space under $q(\tau) \rightarrow q(\tau) + dq(\tau)$. The explicit form of this variation follows from (5.25): for small $d\tau$,

$$dI^a = f_c^{\ ab}[A_{b\mu}(\tau, x)\,dx^\mu + \phi_b(\tau, x)d\tau]I^c, \qquad (5.26)$$

where we have allowed an explicit τ-dependence for the gauge field, and have included a Lorentz scalar gauge field ϕ_a, in analogy with the Abelian case. The quantity $\mathbf{M} = M_a I^a$ undergoes the variation of the path

$$(\tau, x) \longrightarrow (\tau + d\tau, x + dx)$$

according to

$$\begin{aligned}
d\mathbf{M} &= (dM_a)I^a + M_a(dI^a) \\
&= \left(\frac{\partial M_a}{\partial \tau}d\tau + \frac{\partial M_a}{\partial x^\mu}dx^\mu + \frac{\partial M_a}{\partial \dot{x}^\mu}d\dot{x}^\mu + \frac{\partial M_a}{\partial \ddot{x}^\mu}d\ddot{x}^\mu\right)I^a \\
&\quad + M_a[f_c^{\ ab} A_{b\mu}\,dx^\mu + \phi_b d\tau]I^c \\
&= \left[\frac{\partial M_a}{\partial \tau} - f_a^{\ bc}\phi_b M_c\right]I^a d\tau + \left[\frac{\partial M_a}{\partial x^\mu} - f_a^{\ bc} A_{b\mu} M_c\right]I^a dx^\mu \\
&\quad + \frac{\partial M_a}{\partial \dot{x}^\mu}I^a d\dot{x}^\mu + \frac{\partial M_a}{\partial \ddot{x}^\mu}I^a d\ddot{x}^\mu \\
&= \mathcal{D}_\tau \mathbf{M}d\tau + \mathcal{D}_\mu \mathbf{M}dx^\mu + \frac{\partial \mathbf{M}}{\partial \dot{x}^\mu}d\dot{x}^\mu + \frac{\partial \mathbf{M}}{\partial \ddot{x}^\mu}d\ddot{x}^\mu
\end{aligned}$$

in which the spacetime part of the covariant derivative $\mathcal{D}_\mu$ has the form

$$(\mathcal{D}_\mu F_{\nu\rho})_a = \partial_\mu F_{a\nu\rho} - f_a{}^{bc} a_{b\mu} F_{c\,\nu\rho}$$

and a similar covariant derivative for the τ component appears which contains ϕ_a.

Now, the entire structure of self-adjoint equations follows with the replacements

$$\frac{\partial}{\partial x^\mu} \longrightarrow \mathcal{D}_\mu \qquad \frac{\partial}{\partial \tau} \longrightarrow \mathcal{D}_\tau,$$

so that the Helmholtz conditions become

$$A_{\mu\nu} = A_{\nu\mu} \qquad\qquad \frac{\partial A_{\mu\nu}}{\partial \dot{x}^\sigma} = \frac{\partial A_{\sigma\nu}}{\partial \dot{x}^\mu} \tag{5.27}$$

$$\frac{D}{D\tau} A_{\mu\nu} = -\frac{1}{2}\left[A_{\mu\sigma} \frac{\partial f^\sigma}{\partial \dot{x}^\nu} + A_{\nu\sigma}\frac{\partial f^\sigma}{\partial \dot{x}^\mu}\right] \tag{5.28}$$

$$\frac{1}{2}\frac{D}{D\tau}\left[A_{\mu\sigma}\frac{\partial f^\sigma}{\partial \dot{x}^\nu} - A_{\nu\sigma}\frac{\partial f^\sigma}{\partial \dot{x}^\mu}\right] = A_{\mu\sigma}\mathcal{D}_\nu f^\sigma - A_{\nu\sigma}\mathcal{D}_\mu f^\sigma, \tag{5.29}$$

where

$$\frac{D}{D\tau} = \mathcal{D}_\tau + \dot{x}^\sigma \mathcal{D}_\sigma + f^\sigma \frac{\partial}{\partial \dot{x}^\sigma}$$

is the total τ derivative subject to

$$\ddot{x}_\mu - f_{a\mu}(\tau, x, \dot{x})I^a = 0.$$

Since Hojman and Shepley's argument relates only to the commutation relations among the coordinates, not to the structure of the forces, their result carries over unchanged.

In flat spacetime, with $A_{\mu\nu} = g_{\mu\nu} = \eta_{\mu\nu} = $ constant, (5.27) is trivially satisfied and (5.28) becomes

$$\frac{\partial f_\mu}{\partial \dot{x}^\nu} + \frac{\partial f_\nu}{\partial \dot{x}^\mu} = 0 \qquad\Longrightarrow\qquad \frac{\partial^2 f_\mu}{\partial \dot{x}^\nu \partial \dot{x}^\lambda} + \frac{\partial^2 f_\nu}{\partial \dot{x}^\mu \partial \dot{x}^\lambda} = 0. \tag{5.30}$$

Recalling the identity (5.17), we may also write (since the metric carries no group indices)

$$\frac{\partial^2 f_\mu}{\partial \dot{x}^\nu \partial \dot{x}^\lambda} - \frac{\partial^2 f_\nu}{\partial \dot{x}^\mu \partial \dot{x}^\lambda} = 0 \qquad \longrightarrow \qquad \frac{\partial^2 f_\mu}{\partial \dot{x}^\nu \partial \dot{x}^\lambda} = 0,$$

and so the most general form of $f_{a\mu}$ is

$$f_{a\mu} = f_{a\mu\nu}(\tau, x)\dot{x}^\nu + g_{a\mu}(\tau, x), \tag{5.31}$$

where (5.30) requires that $f_{a\mu\nu} + f_{a\nu\mu} = 0$. Finally, applying (5.29) leads to

$$\frac{1}{2}\frac{D}{D\tau}\left[\frac{\partial f_\mu}{\partial \dot{x}^\nu} - \frac{\partial f_\nu}{\partial \dot{x}^\mu}\right] = \mathcal{D}_\nu f_\mu - \mathcal{D}_\mu f_\nu$$

$$\frac{1}{2}\frac{D}{D\tau}[f_{a\mu\nu} - f_{a\nu\mu}] = \mathcal{D}_\nu f_{a\mu\lambda}\dot{x}^\lambda + \mathcal{D}_\nu g_{a\mu} - \mathcal{D}_\mu f_{a\nu\lambda}\dot{x}^\lambda + \mathcal{D}_\mu g_{a\nu}$$

$$(\mathcal{D}_\tau + \dot{x}^\lambda \mathcal{D}_\lambda)f_{a\mu\nu} = \dot{x}^\lambda(\mathcal{D}_\nu f_{a\mu\lambda} - \mathcal{D}_\mu f_{a\nu\lambda}) + \mathcal{D}_\nu g_{a\mu} - \mathcal{D}_\mu g_{a\nu}$$

and since $\dot{x}^\mu$ is arbitrary, we obtain

$$\mathcal{D}_\lambda f_{a\mu\nu} + \mathcal{D}_\mu f_{a\nu\lambda} + \mathcal{D}_\nu f_{a\lambda\mu} = 0$$
$$\mathcal{D}_\tau f_{a\mu\nu} + \mathcal{D}_\mu g_{a\nu} - \mathcal{D}_\nu g_{a\mu} = 0$$

for the fields $f_{a\mu\nu}$ and $g_{a\mu}$. Now, in analogy to the Abelian case, we may write

$$L = \frac{1}{2}M \ \dot{x}^\mu \dot{x}_\mu + A_{a\mu}(\tau, x)I^a(\tau)\dot{x}^\mu + \phi_a(\tau, x)I^a(\tau)$$

and applying the Euler-Lagrange equations, we obtain

$$\frac{d}{d\tau}\left[m\dot{x}_\mu + A_{a\mu}I^a\right] = \frac{\partial}{\partial x^\mu}[A_{a\nu}I^a\dot{x}^\nu + \phi_a I^a]$$

$$M\ddot{x}_\mu + \frac{\partial A_{a\mu}}{\partial\tau}I^a + \frac{\partial A_{a\mu}}{\partial x^\nu}\dot{x}^\nu I^a + A_{a\mu}\dot{I}^a = \frac{\partial A_{a\nu}}{\partial x^\mu}\dot{x}^\nu I^a + \frac{\partial\phi_a}{\partial x^\mu}I^a.$$

Rearranging terms and using (5.26) to express $\dot{I}^a$, we find

$$M\ddot{x}_\mu = \left[\left(\frac{\partial A_{a\nu}}{\partial x^\mu}\dot{x}^\nu - \frac{\partial A_{a\mu}}{\partial x^\nu}\dot{x}^\nu\right)I^a - A_{a\mu}\dot{I}^a\right] + \frac{\partial\phi_a}{\partial x^\mu}I^a - \frac{\partial A_{a\mu}}{\partial\tau}I^a$$

$$= \left[\left(\frac{\partial A_{a\nu}}{\partial x^\mu}\dot{x}^\nu - \frac{\partial A_{a\mu}}{\partial x^\nu}\dot{x}^\nu\right)I^a - A_{a\mu}f_c^{\ ab}(A_{b\nu}\dot{x}^\nu + \phi_b)I^c\right] + \frac{\partial\phi_a}{\partial x^\mu}I^a - \frac{\partial A_{a\mu}}{\partial\tau}I^a$$

$$= \left[\frac{\partial A_{a\nu}}{\partial x^\mu} - \frac{\partial A_{a\mu}}{\partial x^\nu} + f_a^{\ bc}A_{b\mu}A_{c\nu}\right]\dot{x}^\nu I^a + \left[\frac{\partial\phi_a}{\partial x^\mu} - \frac{\partial A_{a\mu}}{\partial\tau} + f_a^{\ bc}A_{b\mu}\phi_c\right]I^a.$$

Comparing this with (5.31), we may express the field strengths in terms of the potentials as

$$f_{\mu\nu} = \left[\frac{\partial A_{a\nu}}{\partial x^\mu} - \frac{\partial A_{a\mu}}{\partial x^\nu} + f_a^{\ bc}A_{b\mu}A_{c\nu}\right]\dot{x}^\nu I^a$$

$$g_\mu = \left[\frac{\partial\phi_a}{\partial x^\mu} - \frac{\partial A_{a\mu}}{\partial\tau} + f_a^{\ bc}A_{b\mu}\phi_c\right]I^a,$$

from which it follows that the field equations are satisfied. Introducing the definitions

$$x^D = \tau \qquad \partial_\tau = \partial_D \qquad f_{\mu D} = -f_{D\mu} = g_\mu,$$

the field equations and Lorentz force assume the form

$$\partial_\alpha f_{\beta\gamma} + \partial_\beta f_{\gamma\alpha} + \partial_\gamma f_{\alpha\beta} = 0$$

$$M\ddot{x}^\mu = f_a^{\mu\nu}\dot{x}_\nu I^a + g_a^\mu I^a = f_a^{\mu\nu}I^a\dot{x}_\nu + f_{a\ D}^\mu I^a\dot{x}^D = f_{\alpha\beta}^\mu\dot{x}^\beta,$$

where

$$f_{\alpha\beta} = \left[\frac{\partial A_{a\beta}}{\partial x^\alpha} - \frac{\partial A_{a\alpha}}{\partial x^\beta} + f_a^{\ bc}A_{b\alpha}A_{c\beta}\right]I^a$$

recovers the usual relationship of the field strength tensor to the non-Abelian potential.

5.3   EVOLUTION OF THE LOCAL METRIC IN CURVED SPACETIME

General relativity has been summarized as: "Space acts on matter, telling it how to move. In turn, matter reacts back on space, telling it how to curve." [11] The action of space on matter is expressed in equations of motion describing geodesic evolution with respect to a local metric $g_{\mu\nu}(x)$. Such equations were found from a Lagrangian in (3.6) and from canonical commutation relations in (5.7). They can also be described in a Hamiltonian formulation on the phase space of position and momentum, an approach amenable to the canonical quantum dynamics for general relativity developed in [12, 13]. To express the action of matter on space, we look to Einstein equations that relate the local metric to sources of mass and energy, which evolve dynamically with τ. We therefore consider a τ-dependent metric that may also evolve along with its sources. One possible approach, proposed by Pitts and Schieve [14, 15], is to develop general relativity on the 5D manifold (x^{μ}, τ), introducing an ADM-type foliation with τ as a preferred time direction. In the approach followed here, we adhere to the restriction imposed in SHP electrodynamics, maintaining the role of τ as external, non-dynamical parameter throughout.

General relativity treats the interval between a pair of instantaneously displaced points in spacetime

$$\delta x^2 = g_{\mu\nu}\delta x^{\mu}\delta x^{\nu} = (x_2 - x_1)^2$$

as an invariance of the manifold. To transform geometry into dynamics, a particle trajectory maps an arbitrary parameter ζ to a continuous sequence of events $x^{\mu}(\zeta)$ in the manifold. For any timelike path we may put $\zeta = s =$ proper time, and although the path consists of instantaneous displacements in a 4D block universe, "motion" is observed through changes in $x^0(s)$ with proper time. Treating the sequence as a function, the invariant interval can be written

$$\delta x^2 = g_{\mu\nu}\delta x^{\mu}\delta x^{\nu} = g_{\mu\nu}\frac{dx^{\mu}}{ds}\frac{dx^{\nu}}{ds}\delta s^2$$

suggesting a dynamical description of the path by the action

$$S = \int ds\, \frac{1}{2}\, g_{\mu\nu}\frac{dx^{\mu}}{ds}\frac{dx^{\nu}}{ds}$$

which removes the constraint $\dot{x}^2 = -c^2$ associated with the usual square root form.

A physical event $x^{\mu}(\tau)$ in SHP theory occurs *at* time τ and chronologically precedes events occurring at subsequent times. The physical picture that emerges in SHP electrodynamics can thus be understood as describing the evolution of a Maxwell–Einstein 4D block universe defined at time τ to an infinitesimally close 4D block universe defined at $\tau + d\tau$. As $c_5 \to 0$, evolution slows to zero, recovering Maxwell theory as an equilibrium limit. The form of the gauge fields draws our attention to idea that while geometric relations on spacetime, such as O(3,1) invariance, are defined within a given block universe, the dynamics operate through the

τ-dependent gauge interaction, and in this sense are defined in the transition from one 4D block manifold to another. We therefore consider the interval

$$dx^{\mu} = \bar{x}^{\mu}(\tau + \delta\tau) - x^{\mu}(\tau)$$

between an event $x^{\mu}(\tau)$ and an event $\bar{x}^{\mu}(\tau + \delta\tau)$ occurring at a displaced spacetime location at a subsequent time, and expand as

$$dx^2 = g_{\mu\nu}\delta x^{\mu}\delta x^{\nu} + g_{5\nu}\delta x^{\nu}\delta x^5 + g_{55}\delta x^5\delta x^5 = g_{\alpha\beta}(x,\tau)\delta x^{\alpha}\delta x^{\beta}$$

referred to the coordinates of x. This interval contains both the geometrical distance δx^{μ} between two neighboring points in one manifold, and the dynamical distance $\delta x^5 = c_5\delta\tau$ between events occurring at two sequential times. This leads to the Lagrangian

$$L = \frac{1}{2}M g_{\alpha\beta}(x,\tau)\dot{x}^{\alpha}\dot{x}^{\beta}$$

and equations of motion

$$0 = \frac{D\dot{x}^{\mu}}{D\tau} = \ddot{x}^{\mu} + \Gamma^{\mu}_{\alpha\beta}\dot{x}^{\alpha}\dot{x}^{\beta} \qquad 0 = \frac{D\dot{x}^5}{D\tau} = \ddot{x}^5 + \Gamma^5_{\alpha\beta}\dot{x}^{\alpha}\dot{x}^{\beta},$$

where $\Gamma^{\gamma}_{\alpha\beta}$ is the standard Christoffel symbol in 5D. But as in the electrodynamic Lagrangian, we do not treat $x^5(\tau) \equiv c_5\tau$ as a dynamical variable, and take the 5-index to denote scalar quantities, not elements of a 5D tensor. This symmetry breaking of 5D $\longrightarrow$ 4D+1 is expressed through the prescription

$$\Gamma^{\mu}_{5\alpha} = \frac{1}{2}g^{\mu\nu}(\partial_5 g_{\nu\alpha} + \partial_{\alpha}g_{\nu 5} - \partial_{\nu}g_{\alpha 5}) \qquad \Gamma^5_{\alpha\beta} \equiv 0 \qquad (5.32)$$

which extends the geodesic Equations (3.6) and (5.7) to 5D.

We define $n(x,\tau)$ to be the number of events (non-thermodynamic dust) per spacetime volume, so that

$$j^{\alpha}(x,\tau) = \rho(x,\tau)\dot{x}^{\alpha}(\tau) = Mn(x,\tau)\dot{x}^{\alpha}(\tau)$$

is the 5-component event current, and

$$\nabla_{\alpha}j^{\alpha} = \frac{\partial j^{\alpha}}{\partial x^{\alpha}} + j^{\gamma}\Gamma^{\alpha}_{\gamma\alpha} = \frac{\partial\rho}{\partial\tau} + \nabla_{\mu}j^{\mu} = 0$$

is the continuity equation. Generalizing the 4D stress-energy-momentum tensor to 5D, the mass-energy-momentum tensor [16, 17] is

$$T^{\alpha\beta} = Mn\dot{x}^{\alpha}\dot{x}^{\beta} = \rho\dot{x}^{\alpha}\dot{x}^{\beta} \longrightarrow \begin{cases} T^{\mu\nu} = Mn\dot{x}^{\mu}\dot{x}^{\nu} = \rho\dot{x}^{\mu}\dot{x}^{\nu} \\ T^{5\beta} = \dot{x}^5\dot{x}^{\beta}\rho = c_5 j^{\beta} \end{cases}$$

combining $T^{\mu\nu}$ with j^α, and is conserved by virtue of the continuity and geodesic equations. The Einstein equations are similarly extended to

$$G_{\alpha\beta} = R_{\alpha\beta} - \frac{1}{2}Rg_{\alpha\beta} = \frac{8\pi G}{c^4}T_{\alpha\beta},$$

where the Ricci tensor $R_{\alpha\beta}$ and scalar R are obtained by contracting indices of the 5D curvature tensor $R^\delta_{\gamma\alpha\beta}$. Since conservation of $T^{\alpha\beta}$ depends on prescription (5.32), we must similarly suppress $\Gamma^5_{\alpha\beta}$ when constructing the Ricci tensor to insure $\nabla_\beta G^{\alpha\beta} = 0$. Working through the algebra we find that $R_{\mu\nu} = \left(R_{\mu\nu}\right)^{4D}$ and obtain

$$R_{\mu 5} = \frac{1}{c_5}\partial_\tau \Gamma^\lambda_{\mu\lambda} - \partial_\lambda \Gamma^\lambda_{\mu 5} + \Gamma^\lambda_{\sigma 5}\Gamma^\sigma_{\mu\lambda} - \Gamma^\lambda_{\sigma\lambda}\Gamma^\sigma_{\mu 5}$$

$$R_{55} = \frac{1}{c_5}\partial_\tau \Gamma^\lambda_{5\lambda} - \partial_\lambda \Gamma^\lambda_{55} + \Gamma^\lambda_{\sigma 5}\Gamma^\sigma_{5\lambda} - \Gamma^\lambda_{\sigma\lambda}\Gamma^\sigma_{55}$$

as new components.

The weak field approximation [11] is generalized to 5D as

$$g_{\alpha\beta} = \eta_{\alpha\beta} + h_{\alpha\beta} \longrightarrow \partial_\gamma g_{\alpha\beta} = \partial_\gamma h_{\alpha\beta} \qquad \left(h_{\alpha\beta}\right)^2 \approx 0$$

leading to

$$R_{\alpha\beta} \simeq \frac{1}{2}\left(\partial_\beta\partial_\gamma h^\gamma_\alpha + \partial_\alpha\partial_\gamma h^\gamma_\beta - \partial_\gamma\partial_\gamma h_{\alpha\beta} - \partial_\alpha\partial_\beta h\right) \qquad R \simeq \eta^{\alpha\beta}R_{\alpha\beta} \qquad h \simeq \eta^{\alpha\beta}h_{\alpha\beta}.$$

Defining $\bar{h}_{\alpha\beta} = h_{\alpha\beta} - \frac{1}{2}\eta_{\alpha\beta}h$, the Einstein equations become

$$\frac{16\pi G}{c^4}T_{\alpha\beta} = \partial_\beta\partial_\gamma \bar{h}^\gamma_\alpha + \partial_\alpha\partial_\gamma \bar{h}^\gamma_\beta - \partial_\gamma\partial_\gamma \bar{h}_{\alpha\beta} - \partial_\alpha\partial_\beta \bar{h}$$

which take the form of a wave equation

$$\frac{16\pi G}{c^4}T_{\alpha\beta} = -\partial^\gamma\partial_\gamma \bar{h}_{\alpha\beta} = -\left(\partial^\mu\partial_\mu + \frac{\eta_{55}}{c_5^2}\partial_\tau^2\right)\bar{h}_{\alpha\beta}$$

after imposing the gauge condition $\partial_\lambda \bar{h}^{\alpha\lambda} = 0$. Using the Green's function $G_{Maxwell}$ from (3.24) for this equation leads to

$$\bar{h}_{\alpha\beta}\left(x, \tau\right) = \frac{4G}{c^4}\int d^3x' \frac{T_{\alpha\beta}\left(t - \frac{|x-x'|}{c}, x', \tau\right)}{|x - x'|}$$

relating the field $\bar{h}_{\alpha\beta}\left(x, \tau\right)$ to the source $T_{\alpha\beta}\left(x, \tau\right)$. In analogy to the Coulomb problem, we take a point source $X = \left(cT(\tau), 0\right)$ in a co-moving frame, with

$$T^{00} = mc^2\dot{T}^2\delta^3\left(\mathbf{x}\right)\varphi\left(t - T\left(\tau\right)\right) \qquad T^{\alpha i} = 0 \qquad T^{55} = \frac{c_5^2}{c^2}T^{00},$$

where $\varphi(\tau)$ is the smoothing function (3.15). Writing $M = m\,\varphi\,(t - T\,(\tau))$ produces

$$\bar{h}^{00}\,(x,\tau) = \frac{4GM}{c^2 R}\dot{T}^2 \qquad \bar{h}^{\alpha i}\,(x,\tau) = 0 \qquad \bar{h}^{55}\,(x,\tau) = \frac{c_5^2}{c^2}\bar{h}^{00}$$

so using $h_{\alpha\beta} = \bar{h}_{\alpha\beta} - \frac{1}{2}\eta_{\alpha\beta}\bar{h}$ and neglecting $c_5^2/c^2 \ll 1$, we see that $h^{00} = \bar{h}^{00}$. Since $g^{\alpha\beta}h_{\beta\gamma} \simeq \eta^{\alpha\beta}h_{\beta\gamma}$ the non-zero Christoffel symbols are

$$\Gamma^{\mu}_{00} = -\frac{1}{2}\eta^{\mu\nu}\partial_\nu h_{00} \qquad\qquad \Gamma^{\mu}_{0i} = \frac{1}{2}\eta^{\mu\nu}\partial_i h_{\nu 0}$$

$$\Gamma^{\mu}_{50} = \frac{1}{2c_5}\eta^{\mu 0}\partial_\tau h_{00} \qquad\qquad \Gamma^{\mu}_{55} = -\frac{1}{2}\eta^{\mu\nu}\partial_\nu h_{55}$$

so the equations of motion split into

$$\ddot{t} = (\partial_\tau h_{00})\,\dot{t} + \dot{\mathbf{x}}\cdot(\nabla h_{00})\,\dot{t}^2 \qquad\qquad \ddot{\mathbf{x}} = \frac{c^2}{2}\,(\nabla h_{00})\,\dot{t}^2.$$

In spherical coordinates, putting $\theta = \pi/2$, the angular and radial equations are

$$2\dot{R}\dot{\phi} + R\ddot{\phi} = 0 \longrightarrow \dot{\phi} = \frac{L}{MR^2} \longrightarrow \ddot{R} - \frac{L^2}{M^2 R^3} = -\frac{GM}{R^2}\dot{t}^2\dot{T}^2$$

and the t equation is

$$\ddot{t} = \frac{2G\,\partial_\tau M}{c^2 R}\dot{t} + \frac{4GM}{c^2 R}\dot{T}\ddot{T}\dot{t} - \frac{2GM}{R^2 c^2}\dot{R}\dot{T}^2 \approx \frac{2GM}{c^2 R}\left(1 + \frac{\alpha\,(\tau)}{2}\right)\dot{\alpha}\,(\tau)\,\dot{t},$$

where we neglect $\dot{R}/c \approx 0$ and $\partial_\tau\varphi \approx 0$ (taking λ large), and define

$$\dot{T} = 1 + \frac{\alpha\,(\tau)}{2} \longrightarrow \dot{T}^2 \simeq 1 + \alpha\,(\tau) \longrightarrow \dot{T}\ddot{T} \simeq \left(1 + \frac{\alpha\,(\tau)}{2}\right)\frac{\dot{\alpha}\,(\tau)}{2}.$$

In the Newtonian case, $\alpha = 0 \longrightarrow \dot{t} = 1$, but this t equation has the solution

$$\dot{t} = \exp\left[\frac{2GM}{c^2 R}\left(\alpha + \frac{1}{4}\alpha^2\right)\right] \longrightarrow \dot{t}^2\dot{T}^2 \simeq 1 + \frac{1}{2}\left(1 + \frac{2GM}{c^2 R}\right)\alpha$$

which, since $2GM/c^2 R \ll 1$, leads finally to the radial equation in the form

$$\frac{d}{d\tau}\left\{\frac{1}{2}\dot{R}^2 + \frac{1}{2}\frac{l^2}{M^2 R^2} - \frac{GM}{R}\left(1 + \frac{1}{2}\alpha\,(\tau)\right)\right\} = \frac{dK}{d\tau} = -\frac{GM}{2R}\frac{d}{d\tau}\alpha\,(\tau).$$

We recognize K on the LHS as the Hamiltonian of the particle moving in this local metric. The mass fluctuation of the point source is seen to induce a fluctuation in the mass of a distant particle through the field $g_{\alpha\beta}\,(x,\tau)$, producing a small modification of Newtonian gravity.

Interactions in SHP electrodynamics form an integrable system in which event evolution generates an instantaneous current defined over spacetime at τ, and in turn, these currents induce τ-dependent fields that act on other events at τ. We expect that in a similar way, a fully developed SHP formulation of general relativity will describe how the instantaneous distribution of mass at τ expressed through $T_{\alpha\beta}(x, \tau)$ induces the local metric $g_{\alpha\beta}(x, \tau)$, which, in turn, determines geodesic equations of motion for any particular event at $x^\mu(\tau)$.

5.4   ZEEMAN AND STARK EFFECTS

As discussed in Section 2.4, reasonable solutions to the relativistic central force problem are obtained in a restricted Minkowski space (RMS) with fixed unit vector n^μ

$$\text{RMS}(n) = x \in \left\{ x \mid [x - (x \cdot n)n]^2 \geq 0 \right\}$$

invariant under O(2,1) but not general Lorentz transformations. Because quantum states are classified by their symmetry representations, Horwitz and Arshansky [18–20] generalized their solutions to the quantum central force problem to an induced representation of O(3,1). Studying the Lorentz transformations on n^μ and the RMS(n), they found the generators $h_{\mu\nu}$ of O(3,1) for the combined space, formed a maximally commuting set of operators, and solved for eigenstates of these operators. The energy levels of these degenerate quantum states split in a constant electromagnetic field—the Zeeman and Stark effects. To couple the electromagnetic field to these states, we construct a gauge theory for the induced representation in its classical form [21, 22].

We denote $\mathring{n} = (0, 0, 0, 1)$ so that the parameterization (2.5) describes RMS($\mathring{n}$). Given the Lorentz transformation $\mathring{n} = \mathcal{L}(n)\, n$ it follows that

$$x \in \text{RMS}(n_\mu) \qquad \text{and} \qquad y = \mathcal{L}(n)\, x \qquad \Longrightarrow \qquad y \in \text{RMS}(\mathring{n}_\mu)$$

and so we may characterize the full spacelike region $x = \mathcal{L}^T(n)y$ by $\zeta = (n, y)$. Since a Lorentz transformation Λ acts as $n \to n' = \Lambda n$ and $x \to x' = \Lambda x$, it follows that

$$x' = \Lambda x = \Lambda \mathcal{L}(n)^T y = \mathcal{L}(\Lambda n)^T \mathcal{L}(\Lambda n)\, \Lambda\, \mathcal{L}(n)^T y = \mathcal{L}(n')^T y'.$$

Thus, y transforms under the O(2,1) little group defined through

$$y \to y' = D^{-1}(\Lambda, n)\, y \qquad\qquad D^{-1}(\Lambda, n) = \mathcal{L}(\Lambda n)\, \Lambda\, \mathcal{L}(n)^T$$

and since $D^{-1}(\Lambda, n)\mathring{n} = \mathring{n}$, the little group preserves RMS($\mathring{n}$). We have taken $\mathcal{L}(n(\tau))$ to be τ-dependent, but one can show that since $d/d\tau$ is Lorentz-invariant and commutes with Λ

$$(\dot{y})' = D^{-1}(\Lambda, n)\dot{y} + \dot{D}^{-1}(\Lambda, n)\, y = \frac{d}{d\tau}[D^{-1}(\Lambda, n)\, y]$$

is form-invariant. Representing the Lorentz transform $\Lambda : x \to x'$ as

$$\Lambda = 1 + \frac{1}{2}\,\omega_{\mu\nu}\,\mathcal{M}^{\mu\nu} + o(\omega^2) \qquad\qquad (\mathcal{M}^{\mu\nu})^{\sigma\lambda} = \eta^{\sigma\mu}\eta^{\lambda\nu} - \eta^{\sigma\nu}\eta^{\lambda\mu}$$

the Lorentz transform $\bar{\Lambda} : \zeta = (n, y) \to \zeta' = (n', y')$ can be represented as

$$\bar{\Lambda} = 1 + \frac{1}{2}\, \omega_{\mu\nu}\, X^{\mu\nu} + o(\omega^2)$$

and the generators are found as

$$X_{\mu\nu} = -\left(x^T \mathcal{M}_{\mu\nu} \nabla_{\mathbf{x}} + n^T \mathcal{M}_{\mu\nu} \nabla_{\mathbf{n}}\right) = -\left(y^T \mathcal{L}\mathcal{M}_{\mu\nu}\mathcal{L}^T \nabla_{\mathbf{y}} + n^T \mathcal{M}_{\mu\nu} D\right),$$

where we introduce

$$S_\mu = \mathcal{L}\,(\partial/\partial n^\mu)\,\mathcal{L}^T \qquad\qquad D_\mu = (\nabla_n)_\mu + y^T S_\mu \nabla_{\mathbf{y}}\,.$$

It is easily shown that for a function of x alone (even as n varies with τ) D_μ acts as a kind of covariant derivative with $D_\mu f(n, y) = D_\mu f(\mathcal{L}(n)^T y) \equiv 0$.

As a classical Lagrangian on the phase space $\{\zeta, \dot{\zeta}\}$ we put

$$\begin{aligned}
L &= \frac{1}{2}M\left(\dot{x}^2 + \rho^2 \dot{n}^2\right) + e\left(\dot{x}\cdot A(x) + \dot{n}\cdot\chi(n)\right) - V(x^2) \\
&= \frac{1}{2}M\left[(\dot{y} + \dot{n}^\sigma S_\sigma y)^2 + \rho^2 \dot{n}^2\right] + e\left[(\dot{y} + \dot{n}^\sigma S_\sigma y)\cdot A^{(n)}(y) + \dot{n}\cdot\chi(n)\right] - V(x^2),
\end{aligned}$$

where ρ is a length scale required because n is a unit vector, $A^{(n)} = \mathcal{L}A$ transforms under the little group, and we used

$$\dot{x} = \mathcal{L}^T \dot{y} + \dot{\mathcal{L}}^T y = \mathcal{L}^T\left(\dot{y} + \mathcal{L}\dot{\mathcal{L}}^T y\right) = \mathcal{L}^T\left(\dot{y} + \dot{n}^\sigma S_\sigma y\right).$$

This L is scalar and represents a generalized Maxwell electrodynamics including n as a new dynamical degree of freedom. The conjugate momenta are found to be

$$\begin{aligned}
p_\mu &= \frac{\partial L}{\partial \dot{y}^\mu} = M\left(\dot{y}_\mu + \dot{n}^\sigma S_\sigma y_\mu + eA^{(n)}\right) \\
\pi_\mu &= \frac{\partial L}{\partial \dot{n}^\mu} = M\left(\rho^2 \dot{n}_\mu - y^T S_\mu p + e\chi\right)
\end{aligned}$$

having used the antisymmetry of the matrices S_μ. The Hamiltonian is obtained from the Lagrangian as

$$K = \dot{y}\cdot p + \dot{n}\cdot\pi - L = \frac{1}{2M}\left(p - eA^{(n)}\right)^2 + \frac{1}{2Md^2}\left(\mathcal{P} - e\chi\right)^2 + V,$$

where $\mathcal{P} = \pi_\sigma + y^T S_\sigma p$. Taking $A^{(n)} = \chi = 0$ and applying Noether's theorem to the variation produced by a Lorentz transformation $\delta\zeta = \frac{1}{2}\omega^{\mu\nu} X_{\mu\nu}\zeta$ we obtain the conserved quantities

$$h_{\mu\nu} = p^\rho X_{\mu\nu} y_\rho + \pi^\rho X_{\mu\nu} n_\rho = y^T\left[\mathcal{L}(n)\mathcal{M}_{\mu\nu}\mathcal{L}^T\right]p + n^T \mathcal{M}_{\mu\nu}\mathcal{P}$$

which satisfy Poisson brackets $\{h_{\mu\nu}, K\} = 0$.

Now interpreting p and π as quantum operators, so that

$$p_\mu = i\hbar \frac{\partial}{\partial y^\mu} \qquad \pi_\mu = i\hbar \frac{\partial}{\partial n^\mu} \qquad \mathcal{P}_\mu = i\hbar D_\mu$$

the $h_{\mu\nu}$ are precisely the Lorentz generators found by Horwitz and Arshansky for solutions $\psi(x, \tau)$ to the Stueckelberg–Schrodinger equation $i\partial_\tau \psi = K\psi$ and satisfy $[h_{\mu\nu}, K] = 0$. This system is invariant under $U(1)$ gauge transformations

$$\psi \to e^{-ie\Theta(\zeta)/\hbar} \psi \qquad A_\mu^{(n)} \longrightarrow A_\mu^{(n)} + \frac{\partial}{\partial y^\mu}\Theta \qquad \chi_\mu \longrightarrow \chi_\mu + D_\mu\Theta.$$

For interactions cyclic in n, we may put $\dot{n} = 0$ for the classical system which remains within $RMS(n)$ with fixed n, so the classical and quantum dynamics reduce to

$$L = L_0 = \frac{1}{2}M\dot{y}^2 - V \qquad K = K_0 = -\frac{\hbar^2}{2M}\frac{\partial}{\partial y^\mu}\frac{\partial}{\partial y_\mu} + V$$

and quantum wavefunctions satisfy $D_\mu\psi = 0$. The Zeeman and Stark effects are thus obtained in perturbation theory by expressing a constant field strength $F^{\mu\nu}$ as

$$A^\mu(x) = -\frac{1}{2}F^{\mu\nu}x_\nu \qquad \chi_\mu(n) = \rho\, A_\mu(n) = -\frac{\rho}{2}\,F^\nu_\sigma\, n^\sigma$$

and writing

$$A_\mu^{(n)}(y) = \mathcal{L}_{\mu\nu}A^\nu(\mathcal{L}^T y) = -\frac{1}{2}(\mathcal{L}F\mathcal{L}^T y)_\mu$$

for the potential in $RMS(\mathring{n})$. To first order in e, the Hamiltonian is just

$$K = K_0 + \frac{e}{4M}F_{\mu\nu}X^{\mu\nu}$$

so that the Zeeman effect follows from $\frac{e}{4m}F_{\alpha\beta}X^{\alpha\beta} \to \frac{e}{4m}F_{ij}X^{ij} = \frac{e}{2m}B^k L_k$ splitting the energy levels along the diagonal component L_k of angular momentum. For the Stark effect, we put $\frac{e}{4m}F_{\alpha\beta}X^{\alpha\beta} \to \frac{e}{2m}F_{0i}X^{0i} = \frac{e}{2m}E^k A_k$, where A_k is a boost, and to reproduce the phenomenology we must include an additional scalar potential $V \to V + A_5$, where $A_5 = -e\epsilon^\mu x_\mu$, hinting at the 5D gauge theory.

5.5   CLASSICAL MECHANICS AND QUANTUM FIELD THEORY

Although quantum field theory differs from classical mechanics in both methodology and results, classical SHP electrodynamics presents a number of interesting qualitative implications for QED.

As seen in Sections 1.3 and 3.1, the Stueckelberg–Schrodinger equation is first-order in τ, and the Hamiltonian operator is a Lorentz scalar, so that manifest covariance is preserved throughout the second quantization procedure. In constructing canonical momenta, the kinetic term for the fields $f_\Phi^{\alpha\beta} f_{\alpha\beta}$ formed from the cross derivatives of a_α leads to momentum fields $\pi_\mu = \partial_\tau a_\mu$ but no π_5 component, because $\partial_\tau a_5$ does not appear in $f_{\alpha\beta}$. In Dirac quantization for gauge theories [23], one inserts a momentum π_5 conjugate to a^5 and a Lagrange multiplier to enforce the primary constraint $\pi_5 = 0$. The secondary constraint (that the primary constraint commutes with the Hamiltonian) leads to the Gauss law $\partial_\mu f^{5\mu} = (ec_5/c)j^5$. But because this system is first-order, one may apply the Jackiw quantization scheme [24], in which we first eliminate the constraint from the Lagrangian by solving the Gauss law, and then construct the Hamiltonian from the unconstrained degrees of freedom, which are the matter fields and the transverse electromagnetic modes. Since the momentum modes are not constrained to be lightlike, as we saw for plane waves in Section 4.4, there can be three transverse polarization modes. The resulting system is amenable to both canonical and path integral quantization.

In canonical quantization, one finds the propagator $G(x, \tau)$ for the matter fields as the vacuum expectation value of τ-ordered operator products (equivalent to a Fourier transform of the momentum representation with a Feynman contour). The propagator enforces τ-retarded causality, with $G(x, \tau) = 0$ for $\tau < 0$, so that SHP quantum field theory is free of matter loops. Extracting the propagator for a sharp mass eigenvalue recovers the Feynman propagator for the Klein–Gordon equation.

As in classical mechanics, quantum systems evolve as τ increases, with advance or retreat of x^0 treated on an equal footing. Perturbation theory is constructed in an interaction picture obtained by a unitary transformation constructed from the scalar interaction Hamiltonian and τ. As a result, this method has been shown [25] to circumvent the Haag no-go theorem [26], summarized as, "Haag's theorem is very inconvenient; it means that the interaction picture exists only if there is no interaction." [27]

As seen in Section 4.7, particles interacting through the electromagnetic field can exchange mass. The treatment of Moller scattering leads to a cross-section identical to the standard QED result for spinless particles when mass exchange is absent. When mass is exchanged, the usual pole in the cross-section at $0°$ splits into a zero and two poles close to but away from the forward beam axis, providing a small experimental signature (and one very difficult to observe).

Because there are no matter loops in this theory, the problem of renormalization reduces to treatment of photon loops in the matter field (gauge and vertex factors become unity by the Ward identities). Mass renormalization can be absorbed into the first order mass term $\psi^* i \partial_\tau \psi$ in the quantum Lagrangian. To remove singularities from the loop contributions to the matter propagator in standard QED, some regularization scheme is required. However in SHP QED, the field interaction kernel (3.11) places a multiplicative factor $\left[1 + (\xi\lambda\kappa)^2\right]^{-1}$ in the photon propagator. This factor acts as a mass cut-off rendering the theory superrenormalizable. Unlike a momentum cut-off, this factor leaves the Lorentz and gauge symmetries of the original theory unaffected, recalling Schwinger's motivation for his "proper time method" discussed in Section 1.3.

5.6 BIBLIOGRAPHY

[1] Dyson, F. J. 1990. *American Journal of Physics*, 58:209–211. https://doi.org/10.1119/1.16188 97

[2] Tanimura, S. 1992. *Annals of Physics*, 220:229–247. http://www.sciencedirect.com/science/article/pii/000349169290362P 97, 107

[3] Hojman, S. A. and Shepley, L. C. 1991. *Journal of Mathematical Physics*, 32:142–146. https://doi.org/10.1063/1.529507 97, 101

[4] Land, M., Shnerb, N., and Horwitz, L. 1995. *Journal of Mathematical Physics*, 36:3263. 97

[5] Santilli, R. M. 1990. *Foundations of Theoretical Mechanics I*, Springer-Verlag. 102

[6] Helmholtz, H. 1887. *Journal für die Reine Angewandte Mathematik*, 100:137. 102

[7] Darboux, G. 1894. *Leçons sur la Théory Générale des Surfaces, 3*, Gauthier-Villars. 102

[8] Dedecker, P. 1950. *Bulletin de l'Académie Royale des Sciences de Belgique Classe des Sciences*, 36:63. 102

[9] Wong, S. K. 1970. *Nuovo Cimento*, 65A:689. 106

[10] Lee, C. R. 1950. *Physics Letters*, 148A:36. 107

[11] Misner, C. W., Thorne, K. S., and Wheeler, J. A. 1973. *Gravitation*, W.H. Freeman and Co., San Francisco, CA. 110, 112

[12] Horwitz, L. P. 2019. *Journal of Physics: Conference Series*, 1239. https://doi.org/10.1088%2F1742-6596%2F1239%2F1%2F012014 110

[13] Horwitz, L. P. 2019. *The European Physical Journal Plus*, 134:313. https://doi.org/10.1140/epjp/i2019-12689-7 110

[14] Pitts, J. B. and Schieve, W. C. 1998. *Foundations of Physics*, 28:1417–1424. https://doi.org/10.1023/A:1018801126703 110

[15] Pitts, J. B. and Schieve, W. C. 2001. *Foundations of Physics*, 31:1083–1104. https://doi.org/10.1023/A:1017578424131 110

[16] Saad, D., Horwitz, L., and Arshansky, R. 1989. *Foundations of Physics*, 19:1125–1149. 111

[17] Land, M. 2019. *Journal of Physics: Conference Series*, 1239. https://doi.org/10.1088%2F1742-6596%2F1239%2F1%2F012005 111

[18] Arshansky, R. and Horwitz, L. 1989. *Journal of Mathematical Physics*, 30:66. 114

[19] Arshansky, R. and Horwitz, L. 1989. *Journal of Mathematical Physics*, 30:380.

[20] Horwitz, L. P. 2015. *Relativistic Quantum Mechanics*, Springer, Dordrecht, Netherlands. 114

[21] Land, M. and Horwitz, L. 1995. *Jounal of Physics A: Mathematical and General*, 28:3289–3304. 114

[22] Land, M. and Horwitz, L. 2001. *Foundations of Physics*, 31:967–991. 114

[23] Dirac, P. 1964. *Lectures on Quantum Mechanics*, Yeshiva University, New York. 117

[24] Jackiw, R. 1993. https://arxiv.org/pdf/hep-th/9306075.pdf 117

[25] Seidewitz, E. 2017. *Foundations of Physics*, 47:355–374. 117

[26] Haag, R. 1955. *Kong. Dan. Vid. Sel. Mat. Fys. Med.*, 29N12:1–37. Philosophical Magazine Series, 746, 376. 117

[27] Streater, R. F. and Wightman, A. S. 1964. *PCT, Spin, Statistics, and All That*, Princeton University Press. 117

Authors' Biographies

MARTIN LAND

Martin Land was born in Brooklyn in 1953. He grew up in the New York City area, strongly influenced by his mother, a social worker who worked with Holocaust survivors, and his father, a second-generation engineer in small manufacturing businesses associated with the garment industry. In his school years he cleaned swimming pools and stables, worked as a carpenter on a construction site, and expedited orders in the garment center. In 1972, he entered Reed College in Portland, Oregon, where he received a Kroll Fellowship for original research which permitted him to devote an extra year to extensive study in the humanities along with his specialization in physics. After completing his BA in 1977, he returned to New York City where he received an M.S. in electrical engineering from Columbia University in 1979 as a member of the Eta Kappa Nu engineering honor society. He joined Bell Laboratories, developing specialized hardware for fiber optic communication with application in computer networks and video transmission. In 1982, he worked as a telecommunications engineer at a major Wall Street bank. Returning to theoretical physics at Hebrew University in Jerusalem, he worked with Eliezer Rabinovicci on supersymmetric quantum mechanics to receive a second M.S. in 1986. In 1985, he married Janet Baumgold, a feminist therapist and co-founder of the Counseling Center for Women. Following a year devoted to full-time fatherhood and another in compulsory national service, he began working toward a Ph.D. in high energy physics with Lawrence Horwitz at Tel Aviv University in 1988. He elaborated many aspects of the classical and quantum theories known as Stueckelberg-Horwitz-Piron (SHP) theory, producing a dissertation developing the SHP quantum field theory. Concurrently with his doctoral work, he was on the research faculty of the Computer Science Department at Hebrew University, developing specialized hardware for parallel computing. After submitting his dissertation in 1995, he taught communications engineering for three years at the Holon Institute of Technology, before joining the Department of Computer Science at Hadassah College in Jerusalem, teaching computer architecture, microprocessors, embedded systems, and computer networking. He was a founding member of the International Association for Relativistic Dynamics (IARD) in 1998 and has served as IARD president since 2006. In parallel to his activities in physics and computer science, he has enjoyed a long collaboration with Jonathan Boyarin of Cornell University in various areas of the humanities, critical theory, and Jewish studies. This collaboration has allowed him to communicate contemporary thinking in physics, especially notions of time associated with SHP theory, to scholars in other fields as modern context for philosophical consideration of temporality.

LAWRENCE P. HORWITZ

Lawrence Paul Horwitz was born in New York City on October 14, 1930. He lived in Westchester County until 1934, then went to London where his father founded and managed a chain of womens wear shops, called the Richard Shops, and then returned to the United States in 1936. After a few years in Brooklyn, NY, his family moved to Forest Hills in Queens, NY, where he learned tennis and attended Forest Hills High School, a school dedicated to teaching students how to think, where he came to love physics. He then went to the College of Engineering, New York University, where he studied Engineering Physics and graduated summa cum laude with a Tau Beta Pi key and the S.F.B. Morse medal for physics. He met a young lady, Ruth Abeles, who arrived from Germany in the U.S. in 1939 and became his wife before moving on to Harvard University in 1952 with a National Science Foundation Fellowship. He received his doctorate at Harvard working under the supervision of Julian Schwinger in 1957. He then worked at the IBM Watson Research Laboratory where he met Herman Goldstine, a former assistant to John von Neumann and, among other things, explored with him octononic and quaternionic Hilbert spaces from both physical and mathematical points of view. He then moved on to the University of Geneva in 1964, becoming involved in scattering theory as well as continuing his studies of hypercomplex systems with L. C. Biedenharn and becoming involved in particle physics with Yuval Neeman at CERN. He became full professor at the University of Denver in 1966–1972; he then accepted a full professorship at Tel Aviv University. After stopping for a year to work with C. Piron at the University of Geneva on the way to Israel, he has been at Tel Aviv University since 1973, with visits at University of Texas at Austin, Ilya Prigogine Center for Statistical Mechanics and Complex Systems in Brussels, and at CERN, ETH (Honggerberg, Zurich), University of Connecticut (Storrs, CT), IHES (Bures-sur-Yvette, Paris), and Institute for Advanced Study (Princeton, NJ), where he was a Member in Natural Sciences, 1993, 1996, 1999, 2003 with short visits in August 1990, and January 1991, working primarily with S. L. Adler. He is now Professor Emeritus at Tel Aviv University, Bar Ilan University, and Ariel University. His major interests are in particle physics, statistical mechanics, mathematical physics, theory of unstable systems, classical and quantum chaos, relativistic quantum mechanics, relativistic many body theory, quantum field theory, general relativity, representations of quantum theory on hypercomplex Hilbert modules, group theory and functional analysis, theories of irreversible quantum evolution, geometrical approach to the study of the stability of classical Hamiltonian systems, and to the dark matter problem, and classical and quantum chaos. He is a member of the American Physical Society (Particle Physics), Swiss Physical Society, European Physical Society, International Association for Mathematical Physics, Israel Physical Society, Israel Mathematics Union, European Mathematical Society, International Quantum Structures Association, Association of Members of the Institute for Advanced Study, and the International Association for Relativistic Dynamics.

Printed in the United States
by Baker & Taylor Publisher Services